Royal Flying Corps Combat Flying Log

The Wartime Story of Reginald Collis, RFC, RCAF

© 2021 Mark Hillier. All rights reserved. No part of this book may be scanned, copied, uploaded or reproduced in any form or by any means, photographically, electronically or mechanically, without written permission from the copyright holder.

ISBN: 978-1-943492-86-2

Book design by designpanache

Elm Grove Publishing
San Antonio, Texas, USA
www.elmgrovepublishing.com

Elm Grove Publishing is a legally registered trade name of Panache Communication Arts, Inc.

Royal Flying Corps Combat Flying Log

The Wartime Story of Reginald Collis, RFC, RCAF

Mark Hillier
with Mick J. Prodger

Also by Mark Hillier

Westhampnett at War

To War in a Spitfire

Joe Roddis: In Support of the Few

Suitcases, Vultures and Spies: From Bomber Command to Special Operations the Story of Wing Commander Thomas Murray DSO, DFC*

A Fighter Command Station at War, A Photographic Record of RAF Westhampnett from the Battle of Britain to D-Day and Beyond

Warbirds, the Diary of a Great War Pilot

RAF Battle of Britain Fighter Pilots Kitbag

Luftwaffe Battle of Britain Fighter Pilots Kitbag

Royal Flying Corps Kit Bag

Thunderbolts Over Burma

RAF Tangmere in 100 items

Also by Mick J. Prodger

Cellini: Freedom Fighter

Luftwaffe vs. RAF Vol. I Flying Clothing of the Air War, 1939-45

Luftwaffe vs. RAF Vol. II Flying Equipment of the Air War, 1939-45

Vintage Flying Helmets: Aviation Headgear Before the Jet Age

The Luftwaffe Guide to Basic Survival at Sea (with Howard Nickel)

Contents

Acknowledgements 7

Introduction 9

Research Methodology, Hints and Tips 16

Reginald Collis 19

Onwards to War! 32

6 Squadron, Abeele 42

24 Squadron, January 1916 58

19 March 1916 61

The Log Book 68

Canada 117

Reginald Collis's Flying Coat 122

Combat Flying Timeline 123

Bibliogrphy & Online Sources 124

Acknowledgements

Thanks primarily to Mick Prodger for helping this project along and publishing it for me. Thanks also to Steve Buster Johnson for giving me the benefit of his research on 6 Squadron and use of photos; to Tangmere Military Aviation museum for letting me use some of their photographs.

To Ian Castle for allowing me to quote from and use information from his website and Andrew Dawrant for his help with accessing the Royal Aero Club Aviators Certificates through The Royal Aero Club Trust. Also, thanks to Serge Comini for use of photos relating to La Gorgue and to Tony Schnurr for the photographs of Collis's flying coat.

Thanks also to Vaughn Hawker and Pen and Sword for allowing me to use extracts from *Hawker VC RFC Ace*.

Introduction

In all of my collecting of RFC items, personal equipment and ephemera, my favourite things are always a pilot's log book. Having flown for the past 30 years, my log book means the world to me, and is such a precious item. I can look back over the years and remember where and when I flew, types etc. Log books tell a story. They demonstrate training, experience, qualifications held and remind you sometimes of the perils of aviation. These days, fortunately, the perils are few and far between compared to the early aviators!

Without doubt a pilot's log book is his most treasured possession. Often the comments in the log tell the story of that man's life, his adventures, toils and troubles, near disasters and for some, the end of their lives. There is nothing more sobering than reading in the log book of a member of aircrew an entry or stamp with "Death Presumed", "Missing in Action" or "PoW" – often seen in WW2 RAF logs. In the first world war it seems the logs just stop, or they take the handwriting of another squadron member or adjutant to complete the entry.

At the outset of the RFC, it was essential that pilots kept a running log of their flights; these books' pages were divided into rows and columns, their format borrowed from ship's logs. A pilot penned on one page the date of a flight, aeroplane used, passenger's name, time in the air and course flown. On the opposite page, he recorded height achieved, distance, weather and remarks. Early on, flights were often limited by the aircraft performance and affected badly by weather and serviceability. The aeroplanes were slow, unreliable and often capable of carrying just an observer and flying only short lengths of time. In 1913, there were no stamps or endorsements indicating that such logs were mandated by authority. The purpose of the logs was to document the capability of the aircraft as much as that of the pilot. Early log books are often small, privately purchased ledger style log books. Later, the army issued and designated the flying log book, Army book 425.

I wrote a book on RFC Kit, and during the process of researching it I had the opportunity to photograph and handle a number of log books. Some were just bound books with drawn sub-divisions; some the formal Army 425. One log book that stood out and grabbed my attention was this very early ledger style log book, which does not seem to have any particular WD marks or Army references. It was in the ownership of a good friend of mine Mick Prodger in the USA who sent me photographs of the log. This particular one belonged to Reginald Collis, who had joined the Royal Flying Corps in 1912 as an Air Mechanic. He was the 109th person to be accepted to the RFC and went on to become a pilot, receiving the 412th Royal Aero Club certificate to be issued.

I had not seen such an early log book before, nor had the opportunity to own one. I knew that it had some entries of encounters with enemy aircraft, but did not know much about the original owner apart from that he was very early into the RFC. I wanted to acquire the log, but the price was beyond reach for me at the time. A few years later I managed to acquire it. It had never left my thoughts – and I knew it would make an interesting topic to research.

I was not disappointed when it arrived, I unwrapped it and eagerly turned the pages of the well-written log. The names and remarks made fascinating reading. It has now occupied five weeks of my life during the COVID

19 lockdown, taking me back to the early days of the RFC, names and places, aircraft and the dangers of WW1 aviation. Collis was not an ace; he was not awarded any gallantry medals or accolades, but he was awarded a King's commission in the field, so he was well thought of by his superiors and colleagues. His ability as a pilot was never recorded, but he survived 128 hours on operations over the Western Front in 1915, which was no mean feat. Not so much because of the risk of being shot down by enemy fire, but more by being hit by anti-aircraft fire, small arms or just mechanical failure.

He flew with, and knew, some interesting aviators and characters during that time. In his 230 hours of service in the air, he had two serious crashes, 20 occurrences of "rough" running engine, valves breaking, big-end damage and other minor damage, two full engine failures, three occurrences of damage caused by enemy action (either "Archie" or return aircraft fire) and eight occurrences of meetings with the enemy in the air. He had claimed in his log book that he "drove down" a number of enemy aircraft – but these are inconclusive.

What a log book! The handwriting is exceptionally clear and well-presented, and just carrying out some basic research though *Ancestry.com* and *Findmypast*, soon revealed that a number of personalities mentioned within the pages were also early to the RFC – and also had fascinating stories.

According to a Canadian newspaper of 1940:

> *He joined the RFC as a technical instructor and a month later was posted to the Central Flying School at Upavon. Shortly after he took his first flight with Air-Vice Marshall Sir Arthur Longmore, then Lieutenant A. Longmore R.N. The next month in September, he served in the first British Army Manoeuvres in which aeroplanes were employed. Although he was still in the ranks the opportunity came to him to take up flying instruction and Sir Arthur saw to it that he had the best, giving him the necessary instruction himself.*

Collis went solo after 2 hours and 5 minutes and soon completed his Royal Aeronautical Club Certificate, being awarded number 411. Collis became a flying instructor and remained at Upavon before being posted to France for frontline duty in May 1915 at the rank of Sergeant. He was at St Omer with the RFC HQ initially, then 16 Squadron, where he was awarded a commission in the field to 2nd Lieutenant in the East Surrey Regiment and seconded to the RFC for his service rendered.

His first squadron posting was to 16 and then to 6 Squadron, where he served until November 1915, flying with none other than Cuthbert Orde, the well-known portrait artist. He also flew with a number of other notable observers, serving alongside such notable early aviators as Captain Lanoe Hawker, VC, DSO as well as Captain Louis Arbon Strange, DSO, OBE, MC, DFC and Bar. He was at the front carrying out daily patrols, artillery observation, reconnaissance, photography and bombing sorties flying a B.E.2c.

Recollections of an Airman by Strange, contains stories of Collis's time on 6 Squadron.

Of the 6 Squadron men with whom he flew, a number had interesting stories: Lieutenant. J. K. R. Howey who, post war, set up the Romney Hythe and Dymchurch railway; Lieutenant E. H. P. Cave, who was subsequently badly injured and sent back to England to recuperate, returned to the front line, and was wounded again in May 1917; and Lieutenant E. G. Bowen, who was wounded twice in action in the space of two months in late 1915, then killed in action as a pilot fighting against Oswald Boelcke of Jasta 2 in September 1916.

For some, the log book interest rests with combat and claims made. Was he an ace? What medals did he get? For me the interest is who did he know and where was he? The human stories that surround the subject.

After a few days I had got to grips with the overall background and career of Collis, but also wanted to look further at what information there was on his engagements and combat with the enemy. My first port of call was to check the RFC Communiqués, which were produced for internal consumption in the air arm and provided comprehensive coverage of observation and bombing sorties, damage inflicted and incurred in clashes with the enemy, intelligence gathered – and more importantly for me, the fact that aircrew are frequently mentioned by name.

Some of his claims and combat highlighted in the log book I found were mentioned in the Communiqués of 1915. There were only three entries for Collis out of the total of eight engagements he had recorded. One aircraft that was definitely confirmed as shot down by Collis was in 1916 whilst he was ferrying an aircraft to France. He shot it down over the Channel on his way to St Omer. It is not easy to summarise his successes in the air, although there are indications that he had witnessed three or four aircraft "out of control" or "driven down", however these engagements were not recorded anywhere else or confirmed. For the time Collis was operating over the front, the claims system is not easy to follow, but books such as *Above the Trenches, A Complete Record of the Fighter Aces and Units of the British Empire Forces 1915-1920* by Christopher Shores, Norman Franks and Russell Guest, explain well what was going on at the time and quantification of claims, which in summary are as follows:

A British or Commonwealth pilot of the Royal Flying Corps and Royal Naval Air Service, or Australian pilots of the Australian Flying Corps could be credited with a victory for destroying an enemy plane, for driving it down out of control, capturing it, or destroying an enemy observation balloon. In the earliest days of aerial combat, in 1915 and 1916, victories could also be awarded for forcing an enemy aircraft to land in either Allied or enemy territory.

The approval system began with an official Combat Report from the Squadron submitted to Wing HQ. They in turn passed the report on to Brigade HQ. Either Wing or Brigade could approve or disapprove it; sometimes one would confirm the victory while the other would not.

Victories were reported by RFC HQ via Communiqué. The deadline for the daily Communiqué (nicknamed "Comic Cuts" by RFC pilots) was 1600 hours (4:00 p.m.). Following a system that did not always report an event on the day it actually occurred added to the confusion caused by dual reporting. Collis has one entry in his log book which appears on another date within the Communiqués which makes it very confusing to work out. What is clear, thanks to his write up of events, is that it is the same combat.

Unlike other air forces of the time, British authorities did not necessarily require independent ground verification of a victory to award credit.

You can of course visit the National Archives and look through surviving Combat Reports for each squadron and read the accounts. No Combat Reports seem to have survived for Collis, so I couldn't add any more to the story of his successful engagements – but he was not a fighter pilot. His principal roles were observation and artillery work, and any engagement would have been a case of survival and, if possible, get the upper hand without being shot down or wounded.

His short career was by no means uninteresting, but it slowly ate away at his confidence and ability to fly. At the end of 1915 he was admitted to hospital for a short convalescent period, although one can only speculate the issues. I suspect that his nerves were frayed and he needed a rest, although to be fair there is no note of this in his records, and I am not suggesting that there is any concrete evidence. Only maybe snippets of information of moving from squadron to squadron and more "engine tests", shorter flights and weather issues may give away the fact that in the end, flying for him was coming to an end.

After his period in hospital in November/December 1915, Collis was posted back to the UK and initially joined 24 Squadron, then forming at Hounslow. This did not last, and his next job was aircraft delivery to the front, flying Farnborough to St Omer.

By 1916, although he was no longer on front line duty he was still actively flying, ferrying aircraft, and it seems that occurrences of engine issues and other problems were not infrequent. There followed a nasty crash into a gasometer at Farnborough, in rough weather, which led to a fractured collarbone, although looking at the photos he was very lucky to get away with that! He was off sick for three months after that event, before returning to flying yet again. Eventually, he notes in his remarks column:

Landed due to feeling ill in the air.

That marks his last operational flight with the RFC, but not his service. On 10 October 1916 he is confirmed as fit only for ground duties.

By the end of 1916, he was posted to the No. 2 School of Aeronautics at Oxford to teach aero engines.

> This information was gleaned from his service record which was found at the National Archives as a digital download. Collis stayed in the RAF, the records are under *Air 76/99*.

In May 1917, Collis was sent to Canada as officer in charge of engine instruction at the No. 4 School of Aeronautics in Toronto. Here he was promoted to Flight Commander on formation of the RAF, and he stayed on post WW1.

Collis was never an ace, never decorated or mentioned in dispatches for his efforts during the war, but he was awarded the 1914-15 Star, Victory and British War medals. Typical of many RFC and RAF pilots during WW1, their exploits only really come to life in their log books and through reading Combat Reports or the RFC Communiqués. The medals themselves are just proof of their time on the frontline. The story behind them is always worth exploring.

He went on to the retired list of the Army Reserve in July 1920 and returned to an automobile engineering firm in Burnley, Lancashire for a period of three years.

Collis could not keep away from aircraft, and in 1923 he returned to Canada to take a commission in the Royal Canadian Air Force. He was given the rank of Flying Officer and posted to Camp Borden. He was the officer in charge of the ground instructional school from 1924 until 1926, when he became Officer Commanding No. 2 Operations Squadron, which he was with until 1927. His next move was to become Workshops Officer at No. 1 Repair Depot, Ottawa. In 1930 he went to Montreal to become a resident A.I.D. inspector at the Canadian Vickers plant.

Subsequently he was posted to RCAF HQ to become a Chief Trade Test Officer, then staff officer training and finally Staff Officer in charge of inspection of aircraft. Upon promotion to Squadron Leader, he was posted to organise and command No. 2 Technical Training School. He was then promoted Wing Commander in April 1938 and became Engineering Staff Officer at No. 1 Training Command HQ in Toronto. He was later promoted Group Captain and remained with the RCAF for the duration of WW2.

His two sons joined up – one being killed in a Catalina flying boat crash, Flying Officer Reginald Collis of 210 Squadron. His other son, Sergeant Raymond Collis, was at No.14 Service Flying Training School in Canada in 1940. He appears to have survived the war.

There was also a daughter, Naomi Collis.

Reginald Collis died in 1969 and is buried in Winnipeg, Manitoba.

> Records for his son, Reginald Collis, were found on *Commonwealth War Graves/ Find War Dead.*

I have the utmost respect for those early aviators who volunteered for service in the air arms, having perhaps only watched an early flying machine from a distance, or if lucky, they might have had a quick flip, but with no real idea of what they were up against. They approached aviation as though it were a big adventure. The pilots lived the high life and were minor celebrities of their time, but paid a high price! Pilots with the RFC gained the nickname "the twenty minuters", referring to the life expectancy of a pilot or observer – and made famous more recently by the *Blackadder* television series.

The truth is somewhat better, although still stark – average life expectancy for RFC aircrew in 1915/19 being 280 hours or so. Life expectancy a few years on, in 1916/1917 was, in reality, more likely to be from three to ten weeks with the statistics for 56 Squadron, as an example, suggesting 41% of those aircrew who started with the unit died within ten weeks; 16% being wounded and 28% taken prisoner, the remainder making it home to possibly help train more pilots. So ther was an 85% chance of death, injury or capture! Not great odds and certainly not a great career option – but compare that to the statistics for those soldiers on the front line!

What is apparent is that many also suffered with mental stresses and torment, now more commonly referred to and recognized as Post Traumatic Stress Disorder. Even after the war finished, many committed suicide or were too ill to work. There is evidence within Collis's log book that he too was suffering with stress or anxiety of some sort.

In my mind the early pilots and observers were like explorers or pioneers, blazing the trail for military aviation against adversity. What makes these men so special is that like the explorer, they had the elements and exhaustion to battle, albeit over short periods of time, but they also had additional risks. Throw in the fact that aircraft and engine serviceability were not great and the risk of being shot down by anti-aircraft fire or by enemy aircraft was high. On top of this, the unpleasant smell and taste of oil over your face from the aircraft engine and being tossed about by the air currents all in all lead to a rather unpleasant experience. For those who mastered the air and made it their element, they had to learn on the job, testing and understanding the limits of their aircraft and gaining new skills.

Group Captain Reginald Collis of the RCAF in 1940
(Photograph found in RCAF wartime publication)

Aerobatics were often an art form to be learned on the job, not taught – and often frowned upon – because if one got it wrong the aircraft would break up or at least be seriously damaged. Indeed, the understanding of materials and how they went together and understanding the stresses and strains on aircraft were in their infancy. Many early aircraft just broke up in flight, overstressed. Indeed, Duncan Grinnell Milne, a pilot of the first war, wrote:

"*...in aviation*", a friend of mine was wont to say, "*there is as much art as science*".

He goes on to say:

...and with aeronautics, in its earlier stages, art often seemed to be marching ahead of science that was in its infancy and waiting for the pilots whose progressive discoveries, be it said, were frequently the result of accident". (Wind in the Wires)

During training, the risks were extremely high, the death toll in training in the UK was around 8,000 by the end of the war. Prior to 1917, when a more formalised method of training was introduced, there were tragically six or seven fatalities a day as well as several injuries. This is even more astonishing when you look at the poor observers; it was even more perilous for them, with no formal training until 1917. Often they were introduced to the aircraft and flying by the pilot they were paired up with. Some observers/gunners were simply spare Air Mechanics who were happy to fly and took to the air with no formal training – and were never recognised with the formal issue of an observer qualification badge.

Collis served his time over the Western Front and was in the thick of the action around the Ypres area. His log is a testament to the bravery of the "average" pilot/observer who never achieved big accolades and would otherwise be lost to history.

I have approached the writing of this book on the basis of a research guide and case study, to try and help anyone starting down to road of researching a relative, or seeking the back story to an item in their collection. The back story behind every item is worth exploring, be it a medal group or a name on a photo.

Collis was an early aviator, and in my view a pioneer, learning on the job. It makes his log book very special, and the aim of this book is to illustrate this man's career and story from his log book and photo collection, to show how much information can be gleaned and stories re-built using primary documentation such as a log book and using the research resources that are available, both in books and online. Hopefully, it will inspire people to look deeper into this type of historical documentation and bring back to life the characters from within.

I hope you enjoy reading about Reginald Collis and his career, as well as the people with whom he flew and got to know during his service.

Mark Hillier,
Fontwell, 2020

The early, ledger-style flying log book of Reginald Collis, RFC. Measuring 8.5" x 13" (21cm x 33cm) it is about twice the size of the later, standardized RFC log books. The log book is unique and is reproduced in full for all to get an insight into the flying career of Reginald Collis. (See page 68).

Research Methodology, Hints and Tips

When I started the research into the log book and archive of Reginald Collis, I knew where to start, which may sound a strange thing to say. When I started doing research into military history and ancestry many years ago, I learned as I went along, not really having anyone to give me any pointers, often finding links to documents or books by chance. With the advent of the computer and world wide web, the ability to research took on a whole different meaning. Today you can enter a name into Google search and often come up with a link or a trail to follow. There is an amazing amount of information out there to find – some good, some bad. Not all the information is completely correct.

So, in starting my research on Collis and his days in the Royal Flying Corps, I thought that I may add some notes on where to look, how I found the information, and some advice for those who may want guidance. Some might say this should all be in the notes section of a book, but I often find that frustrating and I lose the thread.

Also, I wanted to illustrate how much information can be gleaned from these documents. Any militaria that is named: medals, log books and even photographs or clothing, can lead to a story. Sometimes exciting, sometimes sad, you never know what you will uncover – and that is part of the fun! Each item takes you there, helps you connect with the history and events. It can be an emotional rollercoaster for those researching family members.

With the log book, most people would appreciate its age and its importance in terms of the early aviation connection but for me it is much more. Open the pages and it is like a pop-up book of history! Names, places, aircraft numbers can all be researched. Finding out the history and service record of the men within – learning about their experiences, all bring another dimension to research. Live their lives, read around the events and topics contained within. It can be many hours of fun, but sometimes you will get a dead end and it can be frustrating, not being able to make a connection.

The first thing to remember is that you must not give up! You do not need to be a historian or an expert to find exciting information – and there is plenty out there!

The internet is fantastic in helping people access family history information and military records. If you are looking to research a Royal Flying Corps man, there is a high probability you will find some mention. The National Archives has a number of digitized records that you can access so you don't always need to visit Kew – although if you get chance its worth a trip to see some of the non-digitized material like some of the RFC combat reports! There's nothing like holding a combat report that was signed by "Mick" Mannock VC, DSO and two Bars, MC and Bar!

However, let us start with the basics. If you have a surname only and no information about a squadron, service number or other details this is a bit tough! I would suggest narrowing the search down before paying for any digitized records online. The same really with *Ancestry, Find My Past* and others, all have military records but just a surname will cause you problems unless it is an unusual one or you have initials. Something like Air Mechanic Smith would be very difficult to pin down unless you have a service number, for example. In this book you will see

The medal index cards can be found at the National Archives digital collection. There are over five million cards, held in series *WO 372*. Most of the cards are for soldiers in the British Army. The collection also includes most British Army officers, Indian Army personnel, British Army nurses, Royal Flying Corps personnel, Royal Naval Division personnel, some civilians. Most of the cards contain information about campaign medals, which were generally awarded to all those who served overseas. However, some of the cards record entitlement to gallantry and long service awards.

sometimes I could not pin down a name as the log book only has a rank and surname, which is quite common, and I could not narrow down a service record to that location or date in the log.

There are great books containing names and service numbers as well as basic information. From the early entries in the RFC to casualties and lists of aces you will find excellent reference books that will help. As I go through each step of my research on this log book, I intend to illustrate the books I use, websites, service records etc.

With regards to my first port of call, if I have a surname and initials for Royal Flying Corps records, I will always start by narrowing the field down, and I look at the fantastic website *http://www.airhistory.org.uk/rfc/index.html* from which you can download a spreadsheet of each surname in alphabetical order. There are thousands of names on this website and the work carried out is fantastic!

It is a great source of information that helps narrow the search. Here you can find information on where to find records and what is out there for each name. It will give you a rank, surname and initials and a list of records that exist for that person, medal records, casualty records, promotions, awards, Royal Aero Club Certificate numbers, service records, etc. If you have initials for the person you are researching, this may be a good starting point. An unusual surname also offers a shortcut. It will help narrow the search and if you do find your man, it will tell you what other records exits and are worth getting. This can then help you focus with your online purchase of records either digital downloads from the National Archives or one of the other family research sites that have military records.

These sites tend to have links to Medal Index Cards and medal rolls that will give you further information on what medals and awards the individual was entitled too. They will also give information on the theatre of war and date of entry to France, for example. Once you have an idea of service number, name and initials, it's worth having a stab at looking for service records. Don't be too disappointed if you don't find a service record, many were destroyed in WW2.

The National Archives has service records for RFC and the RAF. If the person you are researching is an officer and stayed on through 1918, they will also have RAF records. These are service records of officers who served in the Royal Air Force during the First World War (1914-1918). This collection in series *AIR 76* consists of the records of over 99,000 men.

The records were created from the inception of the RAF in April 1918. However, they include retrospective details of earlier service in the Royal Flying Corps or Royal Naval Air Service, where appropriate. There are also records of RFC airmen held on other family history sites like *Find My Past*. The National Archives research guides are really helpful and self-explanatory. If you're looking for the records of an airman or NCO go to *https://www.nationalarchives.gov.uk/help-with-your-research/research-guides/airman-royal-flying-corps/* and read the advice given.

If you do find the service records and are able to download them, this will give you some good information on dates of enlistment, postings, promotions, trades, next of kin, medical information, awards, squadrons etc. If you have a squadron that say, a pilot or observer served on, you can always try squadron websites. There may be a historian who has studied the records and history of the squadron and may be able to help with your research.

Otherwise, you can try just doing a Google search on names along with 'Royal Flying Corps'. Sometimes on forums, articles, Imperial War Museum memories project pages and such like, you may be lucky enough to hit upon a mention. If your man was a casualty during the war, a search on the Commonwealth War Graves website is worthwhile.

So where did I start with my log book research for Reginald Collis? I had a name from the log, and I knew he was a Sergeant pilot at the beginning of WW1. Fortunately, I also had a newspaper clipping giving me a potted history of his service. A straightforward Google search brought up quite a bit of information straight away. The *Air History* website I mentioned showed he was listed as being in the Royal Flying Corps, so I downloaded the information under C to see what records existed. Luckily, there are only two Collis entries with the initial R., one with R.G., so I knew I had the correct entry. This website told me of his postings, his original regiment, entry number to the RFC as 109, squadrons he was on including 16 and 6, and promotions, which gave me a great starting point. I had a service number too, which was useful as this was not recorded in his log book.

I have used all of the resources mentioned above to collate a picture of Collis and his service career. I have also had some fun researching the types of aircraft he flew, and learned about the differences in performance. I purchased some WW1 trench maps for areas that Collis was flying in 1915 and spent some time plotting his routes on the maps. Also, I have read up the background on the ground war and what the RFC was up to at the relevant points in the war when he was flying over the trenches.

Whether it is a medal group with just a name and number or a flying log book, you can bring these pieces of history to life and get so engrossed in living the history. The information is out there, and you can have some fun delving into the past.

On a serious note, we also owe these brave men and women a great debt, and we should ensure that their names and contribution to the war effort are not forgotten. Behind every name on every grave there is a mass of history and a life that we can bring back and learn about. So many never made it back. Many were never the same again after they returned home. Many never talked about their experiences to their families. I hope by sharing my story it may encourage you to have a go. Find a story. Remember that relative.

Throughout the text I have added notes where I think the reader may want to know where to look for such information.

There are, of course, lots of other resources and books that will help with research and the National Archives online research guides are a great starting point. I have also kept any notes and references on each page so that you do not need to flick backwards and forwards to find the source of the information included or quoted from. The point of this is to give a case study which may help point you in the right direction as well as a fascinating read and an insight into one man's war that you may enjoy.

Reginald Collis

Reginald Collis was born in Dorking, Surrey in 1893, although his parents soon re-located and as a result Reginald was schooled at Burton-on-Trent at Christ School, Rotherham. He was indentured as an apprentice with an automobile engineering company, Jackson Brothers in Blackpool. With an interest in all things mechanical, he soon looked to aviation and became a member of the Lancashire Aero Club in 1909.

The Lancashire Aero Club is the oldest established flying club in the United Kingdom, founded in 1909 to organise the Blackpool Aviation Week, Britain's first officially recognised air show. As a member of this organisation, he serviced the Bleriot monoplane of a Mr. J. Lumb, an early aviator.

A Bleriot Monoplane belonging to Mr. Lumb that Collis looked after at Blackpool. Collis gained his experience of working on aircraft before he joined the RFC, and this would have stood him out from the other trades no doubt and may be why he was so well thought of. *(Mark Hillier Collection)*

After completing his apprenticeship, due to his increasing interest in aviation, he decided to join the newly forming Royal Flying Corps as he had previous experience with engines and airframes. He was attested on 26 June 1912, only two months after the RFC had been formed. His first posting was to Farnborough, no doubt for basic training and to get his uniform.

All of this information was found through period newspaper clippings, many of which can be found online. Websites such as www.newspaperarchive.com and www.britishnewspaperarchive.co.uk amongst others will often find a mention of people, clubs, events etc.

Prior to the inauguration of the Royal Flying Corps, the Royal Engineers had developed an Air Battalion which operated with balloons and the first fixed wing aircraft. On the 13 April 1912, the King Issued a Royal Warrant for a new service and the battalion was replaced with the Royal Flying Corps, which had both a military wing and a naval wing with a joint Central Flying School. Reginald Collis was one of the very early entries into the RFC just after its formation.

According to his service records, Reginald Collis was attested into the Royal Flying Corps on 26 June 1912 and was initially at South Farnborough before being posted to the Central Flying School, formed at Upavon on 19 June 1912, for the purpose of producing professional war pilots. It was initially equipped with eight French Henri and Maurice Farman aircraft fitted with Gnome and Renault engines.

The engines on these early machines were cantankerous (to say the least!) and control over them rudimentary, either by using a fuel lever or blip switch to cut the engine. To reduce lift, you would kill the engine and put the aircraft into a glide. Later rotary engines did have a throttle, but the pilot also had to fine tune the fuel air mixture with a "fine adjustment" lever. A blip switch was still used for landing. The rotary engines on some aircraft also made them more difficult to operate due to the gyroscopic forces produced.

The first course of tuition commenced on 17 August 1912, and the list of notable pupils reads like a who's who of future military aviation. Students included: Hugh Trenchard, who headed up the RFC in the field during the First World War; J. M. Salmond, who later became chief of the air staff; and Robert Smith-Barry, the man who was to introduce a more formal pilot training scheme that has not changed much over the years. Collis was rubbing shoulders with all of them and would have seen them daily. As Collis was an engine expert its likely he was posted to help on ground instruction and maintenance.

Reginald was selected for flying training and took his first trip on 13 January 1913, in a Maurice Farman S7. In looking at what was occurring at the time, CFS had only run one course up to the end of 1912 and Collis started flying instruction before the start of the official second course, which commenced on the 17 January 1913. The men selected for the second course at CFS consisted of five naval, four R.N.V.R., 15 military and six Territorial or Reserve officers, as well as 20 naval ratings, 40 military NCOs and enlisted men for ground school purposes.

Why did Collis get to be one of the first Air Mechanics to learn to fly at Upavon? The press cuttings included within the log book group suggests the answer might be that Collis had been engaged for duty in the RFC as part of the Army manoeuvres in September 1912, when two aeroplane flights were engaged for reconnaissance duties. Lieutenant Arthur Longmore and Hugh Trenchard were manning one of the aircraft. Collis was no doubt involved in maintenance on these manoeuvres and got to know Longmore, who then took him under his wing for flying instruction. I have no evidence of this, but Collis is given

To find out what early instruction was like and how flying training was set up in the early days it's worth downloading a copy of *Earning Their Wings: British Pilot Training, 1912-1918*; a thesis submitted to the College of Graduate Studies and Research on Partial Fulfilment of the Requirements for the Degree of Master of Arts in the Department of History, University of Saskatchewan, Saskatoon, Saskatchewan, Canada by Robert M. Morley at http://citeseerx.ist.psu.edu/viewdoc/download?doi=10.1.1.864.4341&rep=rep1&type=pdf.

flying training out of context with what was occurring at CFS at the time. It may just have been Reginald's engineering prowess and ability that brought him to the attentions of Lieutenant Longmore, the Commander of B Flight at CFS who helped him achieve his goal of becoming a pilot. Longmore records in his book that:

> *It was my flight which started the experiment of training the naval petty officer and army NCO to fly. I see from my records that, out of a total of twelve, nine qualified for the Royal Aero Club Certificate.*
> (*From Sea to Sky*, Arthur Longmore, Geoffrey Bles; First Edition, 1 Jan. 1946)

After Collis learned to fly, there were then a further 20 NCO pilots and men who took a flying course in 1913, of which 14 passed as 2nd Class Pilots. Collis therefore appears to be one of the first of the non-commissioned men to be trained at CFS as there were no Air Mechanics or even NCOs on the first course in 1912. Maybe he was the "experiment" that Longmore mentions?

Reginald was given only the 14th Certificate issued to a non-commissioned officer. He was only the third Air Mechanic of the RFC to be awarded his certificate, the first being given to Air Mechanic William Strugnell (later Group Captain William Victor Strugnell, MC and Bar) Certificate number 253 and the second to Air Mechanic William McCudden, brother of James McCudden, VC (later killed in a flying accident as a flying instructor).

The early years of CFS are covered in a book *C.F.S Birthplace of Air Power* by John W.R Taylor which gives information about the setting up of the School, eyewitness accounts of the location and the accommodation as well as the syllabus at the time which makes for great background reading. With regards list of pilots awarded an aviators certificate you can also finds a free reference by using https://en.wikipedia.org/wiki/List_of_pilots_awarded_an_Aviator%27s_Certificate_by_the_Royal_Aero_Club_in_1912. Here you can get a name and certificate number but not any further details and you would need to look at *Ancestry.co.uk* for photographs or further information.

On 13 January 1913, it was Lieutenant Arthur Longmore, later to become an Air Chief Marshall A. Longmore KCB, DSO who took Reginald under his wing and taught him how to fly. On the first course at CFS in 1912, Arthur Longmore had instructed none other than Major Hugh Trenchard, later Marshall of the Royal Air Force Hugh Montague Trenchard, 1st Viscount Trenchard, GCB, OM, GCVO, DSO.

Arthur Longmore carried out three instructional sorties with Collis in Maurice Farman number 403 on 13, 14 and 17 of January 1913, giving a total instructional time of 2 hours 5 minutes before he allowed Collis off on a solo on 17 January. After 5 hours and 14 minutes flying, he gained Royal Aero Club Aviators Certificate number 412.

Early pilot training was carried out on very basic machines such as the Maurice Farman, with a 70hp Renault Engine. Pilots typically only got 2-3 hours dual instruction before being expected to go solo. In the main, the first training flights occurred with the instructor flying the aircraft while the pupil observed his inputs to the controls and how the aircraft responded to them. When the instructor felt the pupil was sufficiently familiar with the controls of the aircraft, he would allow the pupil to have a go!

Cecil Lewis was one of those pilots with very few hours as he recalls:

> *...one and a half hours dual stood to my credit. I had trundled round the aerodrome with Sergeant Yates my instructor, doing left hand circuits, and made a few indifferent landings.*

At this point his instructor was happy for the student to go solo in the afternoon, if the wind was right!

Initially concerned more with ballooning, as opposed to heavier-than-air flight, the Aero Club was

Period books can give a sense of the type of training environment and first-hand accounts – useful when building up a history of the man you are researching. In this case a well known book on WW1 aviation *Sagittarius Rising* by Cecil Lewis, Peter Davies, London, 1938.

A fantastic photo of the Officers First Course at Central Flying School at Upavon in 1912. In the front row, third from the left is Lieutenant Arthur Longmore who was to train Reginald Collis in January 1913. This photo also has Major Hugh Trenchard on the far right of the middle row who flew with Longmore on the Army exercises of 1912. It is highly probable that Collis helped work on the aircraft flown by Longmore and hence how he managed to get to be trained to fly prior to the start of the 2nd CFS course in 1913. Note the mixture of Army, Navy and RFC uniforms. *(Graham Turner Collection)*

founded in 1901. By the time that the club was granted its Royal prefix on 15 February 1910, it had developed into more of a regulatory role. Indeed, from that year the Royal Aero Club (RAeC) was responsible for issuing Aviator's Certificates, these being internationally recognised under the Fédération Aéronautique Internationale, of which the Club was the UK representative.

The importance of the RAeC following the outbreak of war in 1914 cannot be over-emphasised. Its members included – and trained – most of Britain's military pilots up to 1915, when military schools became established.

By the Armistice in 1918 more than 6,300 military pilots had obtained RAeC Aviator's Certificates. Such was the importance attached to these certificates that until as late as 1917 it remained a requirement that a pilot had to hold a RAeC Aviator's Certificate prior to being granted a commission in the Royal Flying Corps or Royal Naval Air Service.

To obtain an Aviator's Certificate, or "ticket" as it was widely known, a prospective pilot was required to complete three separate, solo, test flights, all of which had to be "vouched for in writing by observers appointed by the Royal Aero Club". The first two "distance flights" needed to be of at least five kilometres, "each in a closed circuit without touching the ground or water". The third test was the "altitude flight" during which "a height of at least 100 metres (328 feet) above the point of departure must be attained; the descent to be made from that height with the motor cut off. The landing must be made in view of the observers."

By passing this examination, the pilot showed that he or she had gained control of an aeroplane efficiently and so could, in theory, take part in all aerial contests and displays run by the Club. It is worth noting that the date listed on an Aviator's Certificate is, generally speaking, not the date on which the pilot actually made the necessary qualifying flights, but the date of the meeting at which the RAeC Committee granted the certificate.

Ancestry.co.uk have the RAeC certificates, three albums of photos of the recipients and the index cards.

Air Mechanic Reginald Collis's instructor, Arthur Longmore, joined the Royal Naval Air Service in 1912. He was one of the first four Naval Officers to be trained as pilots. He was posted to CFS Upavon as an Instructor and taught on the first CFS course in August 1912. He had also taken part in the 1912 Army manoeuvres in a Maurice Farman with Trenchard as his observer. His comments on the use of the type during the exercise, from his autobiography, are worth mentioning as they show both the fragility and the ease of fixing such types. Whilst carrying out observation of troop movements they sought to find some staff officers on a road to whom they could report the advance – which meant landing in an unprepared field:

> *I landed in a stubble field alongside the road and Trenchard handed him a message less than an hour from the time the enemy had been located. Getting out of that field, however, I slightly chipped the airscrew on one of those stick erections which gamekeepers use to stop poachers dragging their nets across a field. Luckily, my ground organization was equal to the occasion, for I had a spare strapped to my private car. One day we flew for over six hours on a tactical reconnaissance, landing and taking off from many of those big grass or stubble fields to be found in that part of the country. On the last landing I broke a longeron, one of the four which carried the tail structure. We found the nearest blacksmith, blew up his fire, and during the night O'Conner made a metal sleeve with which we fixed that longeron. The old Longhorn was once again serviceable by dawn the next day.*
>
> *From Sea to Sky*, Arthur Longmore

Although draggy and ungainly, it seems that the Maurice Farman was not a bad crate and took some punishment, but could also easily be field repaired thanks to the tradesmen of the time.

Arthur Longmore was later to be appointed commander of the seaplane base at Cromarty, and then the experimental seaplane establishment at Calshot. During the war he served as Officer Commanding No. 3 Squadron RNAS and No. 1 Squadron RNAS. After his flying duties he transferred back to sea as an officer on the battlecruiser HMS *Tiger* in 1916.

During his time aboard HMS *Tiger*, he took part in the Battle of Jutland. He obtained a permanent commission in the Royal Air Force in 1920 and was appointed Air Officer Commanding No. 3 Group later that year, before being given command of the RAF Depot in 1921. He was made Air Officer Commanding No. 7 Group in 1924, Director of Equipment at the Air Ministry in 1925 and Chief Staff Officer at Headquarters Inland Area in 1929. Subsequent appointments included Commandant of the Royal Air Force College Cranwell in December 1929, Air Officer Commanding Inland Area in 1933 and Air Officer Commanding Coastal Area (which was renamed RAF Coastal Command under his leadership) in 1934. He went on to be Commandant of the Imperial Defence College in 1936.

The outbreak of the Second World War found Longmore with the rank of Air Chief Marshal and in charge of RAF Training Command. On 2 April 1940, he was appointed Air Officer Commanding in the Middle East. His last role before his formal retirement in 1942 was as Inspector-General of the RAF.

ACM Sir Arthur Murray Longmore, GCB, DSO, DL who was awarded RAeC certificate number 72 when a Lieutenant in the Royal Navy. He taught Hugh Trenchard on the first CFS course and then Reginald Collis at Upavon in January of 1913 prior to the start of the second CFS course. This photograph is taken from the photograph albums of Aviator Certificate holders which were awarded by the Royal Aero Club, now held by the Royal Aero Club Trust, which has kindly given permission to use these images. They can also be accessed via *Ancestry.co.uk* *(Royal Aero Club Trust)*

Maurice Farman S7 of the type flown by Collis on his first solo. *(Mark Hillier Collection)*

For Collis, part of gaining his certificate was the need to carry a passenger, and his first "victim" was a chap called Lieutenant Kennedy on 1 February, 1913 for a short trip of just 13 minutes duration.

Part of the reason for my research into the log was to see how many of the passengers, students and observers who flew with Collis that I could identify. For the NCOs and Air Mechanics, this meant obtaining a copy of *A Contemptible Little Flying Corps* by I. McInnes and J. V. Webb. This publication is great and gives you all the early entry number Air Mechanic and NCOs with service numbers 1 to 1400. With each is a brief description of their service and achievements. If they learnt to fly, the RAeC Certificate number will be mentioned and then you can often get a photo of the recipient. The book does carry images of some of the RFC men when available. As Collis flew with a number of early entry personnel, this made life much easier with identification. With the officers, this was not so easy as it meant looking through several service records of officers with similar surnames and early entry dates to the RFC, to see if I could get a match. I firstly looked up on the *Air History* site the surname of Kennedy, of which there are a number, but only one candidate that was at Upavon at the time that Collis was there. Comparing dates and locations of his RAeC certificate confirmed that this was the correct man. I also managed to find his photograph on *Ancestry* which adds to the story.

Wikipedia can be a great source of information when researching and cross referencing. Early aviators, decorated pilots, events, aircraft, places can all be found and Longmore has his own page *https://en.wikipedia.org/wiki/Arthur_Longmore*. It is always worth comparing sources, written texts etc. Longmore also wrote a book, *From Sea to Sky*, Geoffrey Bles, London, 1946. Books by contemporary pilots and observers can make for fascinating reading as you can learn about the squadron, its personnel, its operations and if you're lucky sometimes a mention of the chap you're researching. However I will add a caution that most WW1 aviation related auto biographies are not cheap, so narrowing down your subject's history and circle of acquaintances would be advisable first before splashing out on books at a fair cost that may have nothing of use.

Collis's first passenger, I am convinced, was James Robert Branch Kennedy, RN, who gained his certificate just a few days after flying with Collis on the 18 February 1913. He gained his certificate whilst flying a Maurice Farman Biplane at Central Flying School Upavon. He was awarded RAeC Certificate number 423 but sadly his flying career was to be short lived, and he was to be killed on 13 June 1913 as passenger of Charles Gordon Bell at Brooklands whilst flying in a Martinsyde monoplane. Sadly, the casualty rate in the RFC became quite high due to training accidents and in this case, Kennedy was one to fall victim to such an accident.

This led me to look at who Kennedy was flying with when he was killed to see what had happened. Charles Gordon Bell was an early British pilot. He was one of the first hundred licensed pilots in the United Kingdom, and in a short career became known as one of the most skilled and experienced pilots in the country. During the First World War he became an ace, shooting down five German aircraft, before returning to England to work as a flying instructor and test pilot. He was killed in July 1918, when an experimental aircraft he was flying crashed in France. Gordon Bell was awarded Aviator's Certificate number 100 by the Royal Aero Club in 1911, making him one of the earliest qualified pilots in the country. As with many early aviators, he had learned at the Brooklands flying school. He then moved to France, where he worked for Robert Esnault-Pelterie's aircraft firm R.E.P.; while demonstrating one of their aircraft to potential buyers in Turkey, he became the first person to fly across the Sea of Marmara.

Over the following years, he became famous in flying circles, and was considered one of the most skilled pilots in the country. By 1914, he had flown over sixty different models of aircraft. However, he had a crash on 13 June with Kennedy who was killed, and Bell was seriously injured, Bell's aircraft crashed whilst flying low over the airfield. Further information on what actually happened is sparce, what is clear is that the accident was ruled entirely due to pilot error, and he was formally cautioned by the Royal Aero Club.

Collis's next passenger flight occurred on the 7 March 1913 in machine number 427 for 16 minutes with Air Mechanic John Roland Gardiner who was RFC number 178. This Air Mechanic was from Portsmouth. He had originally joined up in the Royal Engineers and was posted to CFS on the 4 July 1912. John also subsequently learned to fly and gained Royal Aero Club Aviators certificate number 707 on 11 December 1913 as a Sergeant Pilot.

Again, the source here for Gordon Bell is *Wikipedia*, but if you are researching aviators you may also want to look at *The Aerodrome* website and they too have an entry for Bell – *http://www.theaerodrome.com/aces/england/bell1.php.*

Above: Lieutenant J. R. B. Kennedy, RN, who was learning to fly at CFS alongside Collis. Although he qualified, he was to be tragically killed in a flying accident. *(Royal Aero Club Trust)*

Above right: Gordon Bell was awarded Aviator's Certificate number 100 by the Royal Aero Club in 1911, making him one of the earliest qualified pilots in the country. *(Royal Aero Club Trust)*

Right: Air Mechanic John Gardiner, who Collis took as a passenger on the 7th of March. *(Royal Aero Club Trust)*

At the outbreak of war, Gardiner was posted to France as one of the first in 3 Squadron and flew over with pilot Lieutenant Joubert de la Ferte. He was later to be decorated with a Meritorious Service Medal in 1916 by Major Lanoe Hawker VC, DSO, the then commanding officer of 24 Squadron which had formed at Hounslow at the beginning of the year. Collis was also to join 24 Squadron for a short while at Hounslow in the beginning of 1916 but was soon posted out for ferry duties. Gardiner was later to serve his country during WW2 at the rank of Wing Commander.

On the 13 March he took up Air Mechanic Lloyd for 16 minutes air experience. Sadly no reference to Lloyd can be found in any records, but the same day he also flew with Air Mechanic Thomas Warren for 23 minutes.

Air Mechanic Thomas Warren, RFC Number 153 who was from Loughborough, had joined up into the Coldstream Guards and later moved to the Royal Engineers, 2 Company Balloon School. He was posted to Central Flying school on 27 June 1912 and like Collis, also learned to fly, gaining his wings on the 26 February 1914 and issued with certificate number 748.

Subsequently he was posted to 6 Squadron on the 6 October 1914, a squadron that Collis was to be posted to in 1915. He fortunately survived the war. Warren appears in Reginald's log book twice with one flight on the 13 March 1913 and then again on the 18 March both flights being in a Maurice Farman.

His next passenger had qualified already for his flying license with the Bristol School at Salisbury. This was Lieutenant H. D. Vernon, RN, Certificate number 404, which he achieved in January 1913. He flew with Collis for 10 minutes in machine number 425 on the 18 March 1913. In his service records it shows he gained his ticket at CFS

Air Mechanic Thomas Warren, who Collis flew with on the 13 March 1913. *(Royal Aero Club Trust)*

Henry Doone Vernon was a pilot in the RNAS and went missing 29 September 1914 whilst on a seaplane patrol in the North Sea. (Royal Aero Club Trust)

in January 1913 and the CFS Commandant makes the comment on his flying ability in his service records in April 1913 as "Keen and Able, a good pilot on BE and MF".

Although in the logbook, Vernon's surname only is mentioned, I did a search on *Ancestry* to start to identify any likely candidates who were flying or learning to fly at that time, at Upavon or close by. This gave me the initials to then check through the facts and start looking for service records. Through a search on the Commonwealth War Graves website, I found Vernon as a casualty. *https://www.cwgc.org/find-war-dead/casualty/3050289/vernon,-henry-doone/* I then managed to locate his service records, which are digitized at the National Archives under file *ADM 273/31/23*. A Google search also brings up a number of references.

Reginald took up Air Mechanic Edward Surman with him in a Maurice Farman on the 18 March 1913 for a 17-minute trip around the local area.

Air Mechanic William Edward Surman was an early RFC entrant with the service number 305. He was originally in the Royal Berkshire regiment, but after joining the RFC he was posted to Central Flying School on the 10 August 1912. He later served in France during 1915 and was to become a Sergeant pilot. He was awarded certificate number 4434 on 2 April 1917. Surman later served in the RAF in 1918/1919.

Sergeant Harold Victor Robbins was the next passenger, later on in the afternoon of the 18 March. He was RFC entry number 263. He hailed from Kenilworth. In time, he too learnt to fly and was awarded RAeC certificate 463 on the 22 April 1913 at CFS Upavon. He served overseas and gained a Mention in Despatches. He was to survive the war and stayed on to serve in the RAF.

Air Mechanic William Edward Surman, RFC number 305, who flew with Collis on the 18 March. *(Royal Aero Club Trust)*

Sergeant Harold Victor Robbins, who flew with Collis in March 1918, gained his wings later the following month. *(Royal Aero Club Trust)*

An interesting book for checking awards such as DSO, MC and MID awards is *Royal Flying Corps (Military Wing) Casualties and Honours During the War of 1914-1917* compiled by Captain G. L. Campbell, RFA, written in 1917 and reprinted in 1987.

The end of the month and early April sees Collis engaged in some early wireless experimentation. His passengers were both Royal Navy, with the first sortie being with Petty Officer Hogan on the 8 April, 1913 in a Maurice Farman. The sortie lasted 50 minutes as they flew from Upavon to Andover and on to Larkhill testing the wireless set, the first of two such sorties on that day. The following day they were airborne for two hours, flying from Salisbury to Grendon Underwood, Buckinghamshire, but a rough landing in a field damaged the aircraft and it had to be recovered by road. The RFC had at this stage of its development set up an experimental branch of the Military Wing, Commanded by Major H. Musgrave, and its tasks included looking at the development of wireless, amongst other things, and it's likely that Collis was assisting with this role.

By 1914, the use of wireless telegraphy was becoming more established, with the *Army Field Artillery Training Manual of 1914* stating, "...employment of aircraft both during the period of tactical reconnaissance which precedes the battle, and in the battle itself, the aircraft may assist the artillery." This assistance can only be offered if there is two-way communication between the aircraft and the batterie. To this end it covers signalling in some depth, going on to say, "...the means of communication from aircraft are, i) Wireless Telegraphy, ii) Visual Signals, iii) Sound Signals, iv) Dropping written messages.

May 1913 saw a change in role for Reginald as an instructor at Upavon, even though he has very few hours himself. He commenced tuition with Air Mechanic A. Turner, his first pupil, on the 27 May 1913. Turner was RFC entry number 594 who had joined up on 10 February 1913. He completed his recruit training at Farnborough. He was subsequently posted to 3 Squadron at RFC Larkhill and served overseas in France. He remained in the service and joined the RAF in 1918.

Reginald went on to instruct throughout May and June 1913, in addition to giving many passenger flights. One of his students is mentioned as Air Mechanic Webb. This appears to have been Walter George Webb, who was number 191 into the RFC. His identification was one purely of elimination from the other Webbs that are mentioned in *A Contemptible Little Flying Corps* – literally by looking at who could and could not have been at CFS at the time of the flight shown in Collis's log. Luckily, there were not many candidates to choose from at this time! He had originally served in the Royal Engineers (specifically No. 1 Balloon Squadron at Farnborough) and he served with James McCudden (later VC) who was at this stage an Air Mechanic himself. Webb is mentioned in McCudden's book *Flying Fury*.

By checking the surname 'Webb" in the early RFC numbers against possible locations in service history, I managed to confirm it was Walter Webb, who is also mentioned in *Flying Fury, Five Years in the Royal Flying Corps* by James T. B. McCudden, on Page 14.

Webb was posted to 1 Squadron on 6 July 1912, then later to 3 Squadron in November 1913. McCudden says of Webb:

> *Webb was also very keen to fly at all times, and when either of us three were on the aerodrome hardly anyone else ever got a flight at all.*

Webb flew to France with 3 Squadron on the 16 August 1914. He later learned to fly and gained Royal Aero Club Certificate number 1130 on 11 June 1915. He was later to be Mentioned in Dispatches on 22 June 1915 as a Sergeant but sadly he was to lose his life whilst in action on the 26 January 1917. He was captain of a Sopwith $1^1/_2$ Strutter, number A1074, when sadly he and his observer, Corporal R. D. Fleming were shot down in flames.

Entries on the loss of Webb and Fleming can be found in both *The Sky Their Battlefield* by Trevor Henshaw and *Airmen Died in the Great War*. Both books give details of the aircraft, squadron and serial numbers. Also, a Commonwealth War Graves search will give a place of burial or mention on a memorial.

By the end of July, Collis had completed just over 25 hours of flying. From July to December, he does not appear to be teaching but the notes in his log indicate testing, spiral descent and glides, completing a further 15 hours, with the longest flight being of 45 minutes and a climb up to 5,200 feet, the greatest height he had achieved to date!

1914 sees Reginald still operating from Upavon, this includes a few cross-countries and a forced landing to boot! The entry in his log for the 16 February stating:

Forced landing, ignition wire broken, engine cut out at 3,000 feet, landed in ploughed field, good landing, water in jet, repaired same, engine ok.

A common entry in many of the early log books, often the engine or aircraft succumbing to failures caused by lack of ignition, fuel or other failure caused by vibration or poor design. A landing in the nearest field being the order of the day and a phone call back to base, a mechanic would be despatched to make good the problem. Luckily, Collis was skilled enough to be able to get out and make his own temporary fix, which he does a number of times throughout his log entries. He had a sympathy for the engines and his knowledge saved him from a long wait a number of times.

On the 28 March 1914, Reg flies with Sergeant Fred Farrer. Another early entry to the RFC with entry number 751, Farrer was originally an armourer Staff Sergeant in the Army Ordnance Corps, but he transferred to the RFC in May 1913. Fred went on to earn aviator certificate number 685 sometime after Collis at Upavon.

Period magazines can be useful for research if you can get hold of copies or find a library which has a set of *Flight* magazine. These often contain information on pilots within the RFC and RAF.

He was subsequently posted to France and served in the rank of Sergeant at the Aircraft Park on the 16 August, 1914. Commissioned as a 2nd Lieutenant on 2 March 1917, he was sadly to be killed on 28 November 1917, whilst flying an R.E.8, no. A4474 at the Artillery and Infantry Co-Operation School in Britain.

March to May 1914 sees Reg still flying from CFS and life is still eventful with an entry for a forced landing on a cross country from Upavon to Hungerford via Swindon on the 8 April. The entry reads:

...forced landing at Hungerford, engine missing badly, water in jet number 5 + 6-cylinder plugs sooted, repaired same ok.

Sergeant Fred Farrer. Collis flew with him for 25 minutes on the 28 March 1914. He was later killed in a flying accident. *(Flight Magazine 13 Dec 1913)*

Yet again Collis was able to look after himself and resolve the situation, making a safe landing back at base.

The rest of his flights for May show he was working up towards his Superior Brevet Test. At this point, it's worth considering the training at that time. Collis had already qualified for his Royal Aero Club Certificate. Army Order 20 of 1912 laid out the expectations of candidates who wished to obtain entry to the new aeronautical service and gain their wings as follows:

> *A gentleman not holding a commission who desires to join the Royal Flying Corps as an officer will forward his application to the commandant Central Flying school, quoting the number of his Royal Aero Club Certificate and stating which wing of the corps he wishes to join. If selected for the Military Wing, he will be granted a commission as 2nd Lieutenant on probation in the Special Reserve of Officers. The training of these officers will normally be the same as that prescribed for officers of the Regular Army, and they will receive under the same conditions the sum of £75 if they have obtained the Royal Aero Clubs Certificate at their own expense.*

It was expected that pilots would qualify privately by their own means for an RAeC Certificate, and they could claim some of that money expended back through public funds, then continue on to complete a Central Flying School Course. To gain one's ticket in 1914, a pilot could – with good weather – complete the training within a week.

The requirements included two five-kilometre flights flown around two posts 500m apart, with the direction of flight being reversed at each turn, so five figures of eight, an altitude gain flight of 100m, and the ability to spot land, engine off, within 50m of a designated spot (1914 rules).

Collis had trained at CFS and was, to start with, designated as a 2nd class flyer. His Superior Brevet Test, the details of which are not completely clear, involved a number of cross-country flights, a climb to height followed by an engine off spiral descent to a spot landing. Collis completed the test satisfactorily.

He completed his flying in August 1914 with a grand total of 55 hours and 43 minutes and is posted to HQ at Farnborough until his departure for France in 1915. Bizarrely he does not carry out any flying in that time.

Collis's service papers, which are available via the National Archives digital downloads, show his postings and date from which the posting is effective. Reference *Air 76/99*.

Onwards to War!

Collis's service records state that he was to report to the Officer Commanding of South Farnborough with effect 5 April 1915, with a view to joining the expeditionary force at once. Reginald arrived in France on the 10 May 1915 and reported to St Omer where he commenced with his first flight on the Maurice Farman Shorthorn, number 1857.

Air 76/99/50 for Collis Service records at the National Archives.

A Maurice Farman Shorthorn M.F.11, the type in which Collis took his first sorties and operational sorties in France in 1915. *(Mark Hillier Collection)*

St Omer was an important airfield for the RFC, and on the 11 September, 1914 the first RFC aeroplane touched ground at the aerodrome. It was on the 8 October 1914 that the Headquarters Royal Flying Corps arrived and took up residence at the aerodrome. It was to here that Collis arrived in May 1915. For the next four years, St Omer was to be a central hub for the RFC. Reginal Collis was to continue flying into St Omer in 1916 whilst ferrying aircraft from Farnborough.

Most squadrons only used St Omer as a transit camp, whilst on their way to other locations, but the importance of the site grew as logistical support became its primary function. Major General Hugh Trenchard held his headquarters here up until the end of March 1916 and it returned again for a few months in 1917.

Collis moved to La Gorgue Airfield, the home of 16 Squadron, on 16 May 1915, although he does not appear posted officially until June.

An aerial view, taken approximately in an easterly direction, of the Beaupré Abbey archaeological site, before the confluence of the river Lawe with the Lys. Roughly the location of La Gorgue airfield. In the centre of the photo, to the right and above the airplane wing, the end of the path leading to the site, then the angled fishpond and its extension which once surrounded the Beaupré farm (reduced here to piles of rubble); the sheepfold, still visible today, faces the remains of the farm. Heading towards the Lys, to the left of the photo, the foundations of three wings of the Abbey Square which were brought to light during archaeological excavations (11/1991–06/1992). The land between the fishpond and the abbey square served as a living space for the staff of the La Gorgue aerodrome/Beaupré during the First World War. (*Photography Via Serge Comini: J. Denoeud, 1992*). Text extracted from the book *1914-1918 - Abbaye Notre-Dame de Beaupré-sur-la-Lys, des Hommes sur les terres et les chemins des Dames*, Serge Comini, September 2018, ISBN: 978–2–9564578–0–0.)

The Abbey farm of the former Cistercian Abbey of Beaupré-sur-la-Lys taken after the First World War. A view that Collis would have been familiar with as the site is adjacent La Gorgue airfield. Beaupré farm was located next to the living quarters of the La Gorgue aerodrome. (via Serge Comini, Photograhy: Coll. J. Coupet. Text extracted from the book, *1914-1918 - Abbaye Notre-Dame de Beaupré-sur-la-Lys, des Hommes sur les terres et les chemins des Dames*, Serge Comini.

Collis flew his first reconnaissance sortie from La Gorgue with Captain G. R. Bradley (of the 4th Cavalry, Indian Army) on the 17 May for a duration of 1 hour and 50 minutes.

Captain Bradley was involved in a combat with an Albatross only a few weeks later, on the 29 May, near Douai, whilst acting as observer for Captain Porter in a Maurice Farman Shorthorn. The combat was inconclusive and neither were injured.

These first few flights in France saw Collis climb to heights that he had never been before. 5,000 feet had been his previous maximum climb, but now most of his reconnaissance sorties would be made at 6,000 to 8,000 feet. It was much colder at that height in an open cockpit, and consideration was required as to getting suitably dressed to stay warm for sorties regularly of two hours and more. Whatever practice he had over Salisbury plain probably did not prepare him for the crucible of war and the realities of operating an aircraft over the front lines.

Information about the combat between Porter and Bradley and the Albatross was sourced from *The Sky Their Battlefield* which confirmed Bradley was on 16 Squadron before his official attachment. It also confirms Collis being attached from the date he flew with Bradley in May.

Bradley is a tough nut to crack in terms of research, I initially used the *Air History* website to get the list of RFC personnel with the surname Bradley as I had no initial to go by and at this stage Collis is not officially attached to 16 squadron but is flying from their airfield at this time. There are a lot of Bradleys listed! He appears as being attached to 16 squadron with effect from 16 June 1915 as an observer. All of this information appears to show that Collis was flying with 16 Squadron from the 17 May 1915 and that paperwork and dates of all of the attached personal seem to get tidied up in June of that year. By looking at all the entries for that time in Collis logbook, it confirms his flying with 16 Squadron and then I could fill in the gaps.

The extreme cold and exposure were an issue and highlighted a shortcoming of the issued RFC kit at altitude, as described by Captain Duncan Grinnell-Milne, a pilot of 16 Squadron, who arrived after Collis had departed to 6 Squadron. He was posted in during September of 1915, and he flew the B.E.2c:

...at 14,000 feet over Northern France in November, one realized how very exposed were the seats in a B.E. I was thoroughly chilled myself and I had a windscreen, whereas the observer's had been removed to make room for the forward gun mounting. With the engine running slowly it was possible to make oneself heard. I shouted to ask how he felt. He turned, trying to grin; but he could barely move his jaw enough to shout something back at me. I only got one word: "awful"; sufficient to make me push the stick forward for a faster descent.

One of the disadvantages of flying so high was that it seemed to take such a long time to come down, especially as I had to run the engine every now and then to make sure it was not getting cold. Had we met an enemy aeroplane during that glide we would have had to run for it; the observer could never have handled the gun with frozen fingers. It did not begin to be noticeably warmer until we were below 6,000 feet, but I went on down and did not flatten out until we were at three thousand. He continues: *My hands and feet were aching from the recent freezing cold and I could imagine how the observer must be suffering from the way he kept bouncing about in his seat, stamping, clapping his hands and rubbing his face. Thawing out is agony, but I was glad to notice that his cheeks were no longer white.*

Wind in the Wires, Duncan Grinnell-Milne MC, DFC

Despite the introduction of fur-lined flying caps as well as the use of layers of clothing, leather overcoats and gloves, the cold at this altitude for any period could be intolerable, and indeed many aircrews suffered frostbite and the onset of hypoxia. If they were sweating after enemy action or the kit had become wet through rain on the climb up, ice started to form next to the skin which was particularly unpleasant. Collis would have to get used to operating at these heights to avoid artillery shells and ground fire.

What follows appears to be a crash course in reconnaissance and photography in May and June. As well as reconnaissance, he was sent off on a bombing expedition on the 25 May carrying six 20lb bombs, although no target is identified.

The look of an early aviator from about 1915/1916, dressed in as much warm clothing as one could find to try and battle the cold at altitude. This pilot stands next to his B.E.2 in an issue long leather flying coat; the first ones were a shorter version of this. He would be wearing thick fur gloves and he has a fur-lined leather flying helmet. Collis flew the B.E.2 with 6 Squadron in action. *(Mark Hillier Collection)*

Reginald also flew with a Lieutenant Gordon Ray Elliot and an Air Mechanic Bush on a number of wireless practice sorties. He flies a reconnaissance sortie to the railway triangle at La Basse with Elliot on 7 June, as well as a practice photography flight the next day. This use of Air Mechanics in the other seat, to operate machine guns or operate wireless, is more widespread than most people are aware and many of these went on to get formal recognition as observers once enough hours had been completed.

The first official course for observers did not commence until 13 July 1914, when ten officers arrived at Netheravon for training. Even when war broke out the RFC did not have enough observers. To start with, the rear seat was often occupied by another pilot out of necessity. In 1914 men were identified specifically to be sent overseas as observers, but again these officers had very little in the way of aerial training, often learning on the job and qualifiying for the role at the front! At this time the role of observer was mainly being filled by commissioned ranks, but some of the first airmen to be killed during WW1 were non-commissioned ranks.

The practice of using other non-commissioned volunteers to fly also seemed to become a regular occurrence, with many later becoming qualified in the field as observers. There was also a book available, published in June 1915, called *RFC Notes for Observers*, which along with lectures on the squadron, went some way to helping define what was expected and to offer advice – but was clearly no substitute for a formal training course. By mid-1915, some observers attached to Reserve Aeroplane Squadrons or service squadrons acting as training units, were getting some training in map reading and other key jobs, such as wireless operation.

Lt Gordon Roy Elliott originally enlisted in the 3rd Dragoon Guards but later transferred to the RFC and learnt to fly gaining certificate number 1872 on 11 October 1915. He is noted as serving at 5 Reserve Aeroplane Squadron in October 1915 and later with 33 Squadron at Patchway. *(Royal Aero Club Trust)*

At this stage in the war, the RFC was using the "Transmitter Type No. 1" on the Western Front for artillery spotting duties. It was first used by the RFC during the battle of Neuve Chapelle in March 1915. It was also used by the British Army in France for artillery cooperation, and in Home Defence for spotting work. Transmitter Type No. 1 was a lightweight, simple spark gap transmitter, assembled into a gas-tight box with its inductance calibrated in wavelengths and inductive coupling provided by a wander plug. The transmitter and Morse key were totally enclosed to prevent the spark igniting petrol vapour in the cockpit. It was usually mounted on a tray on the side

information sourced at *http://marconiheritage.org/ww1-air.html*, an excellent resource on wireless work.

of an aircraft's fuselage, and the equipment required a complete overhaul after every flight.

The action heated up on his first trip with Major Edmund William Furse on 15 June 1915, whilst on a Wireless Reconnaissance at 8,000 feet:

> *Wireless reconnaissance [Lens] 'Archie' doing good shooting, large pieces of shell embedded in right hand short undercarriage strut, section of left-hand top tail boom shot away, top main plane and tail plane damaged, machine 1857 discarded.*

Captain Edmund William Furse. *(Royal Aero Club Trust)*

It is likely that Collis had come across Furse previously in his time at CFS, as Captain Furse was attending staff college in 1913 when he found himself on the 1913 War Games in September of that year. He was one of the first non-pilots to fly as part of an practice sortie with the RFC.

Captain Edmund William Furse became a staff officer at St Omer in 1915 after flying as an observer in the first year of the war. Collis was to fly with him on a number of occasions.

Captain E. W. Furse was one of the first officers to be sent to the BEF as a replacement for the early RFC casualties, being requested on 18 August 1914. He was a Royal Artillery officer and a Staff College student, nominated to join the RFC in the event of mobilisation. He learned to fly and gained his certificate number 1706 in December 1915.

He is mentioned in Volume I of *The War in the Air* due to a flight on 31 August 1914, when he was observer to Lieutenant A. E. Borton of No. 5 Squadron RFC, when they were the first to detect the south-east swing of von Kluck's Army.

The War in the Air; Vol. 1 by Sir Walter Alexander Raleigh Published by Edward Arnold & Co., London. (1922).

Furse also flew quite a bit as an observer to Louis Arbon Strange, DSO, OBE, MC, DFC and Bar during the months of September and October 1914. In his book *Recollections of an Airman*, Strange states:

> *I did a lot of artillery observation work in those days, my observer generally being Capt. Furse who always used to do two shoots a day with his own battery.*

Furse was heavily involved in trying to liaise with the artillery, and men on the ground, as can be seen from the entry in *Flying Corps Headquarters 1914-1918* by Wing Commander the Hon. Maurice Baring, OBE, published by G. Bell and Sons, London (1920):

> *One evening in April [1915], just as we were finishing dinner, one of the motor cyclists flung open the dining room door and said, in breathless dramatic tones; "Sir, a Zeppelin has been reported flying towards St Omer" We went up to the aerodrome, but save for a display of searchlights nothing happened. Furse – a gunner who had been*

wounded as an observer early in the war, came to us as a Staff Officer. He understood the possibilities of aircraft and artillery from both sides, and what should be the nature of their co-operation. On April 7th I went with him to see various battery commanders. We found the gunners extraordinarily sticky with regard the co-operation of aircraft with artillery. They seemed to have no belief in it at all. And all Furse's arguments fell on deaf ears. It is intensely trying to have to deal in war time with a new weapon.

Whilst attached to 3 Squadron, Furse also went up to the front lines to send signals to aircraft that could then return messages to the Advanced HQs; his involvement is also mentioned also in McCudden's book *Flying Fury*, Bailey Brothers and Swinfen Ltd., 1973, page 75:

…Major E.W. Furse, accompanied by Flight Sergeant W. Burns of No. 3 Squadron, went out with the advancing infantry and signalled by lamp to an aeroplane which then flew back and dropped messages at advanced wing headquarters. There is a record of three messages, one on the 25th (September), and another on the 27th, and the last on the 28th, which had reference to the attack by the Guards on Pit 14 Bis. Flight Sergeant Burns, soon after this last message was sent, was hit in the head by shrapnel and died of his wounds. Major Furse, however, carried on, making signalling arrangements from a forward position. The experience of battle showed that a more elaborate system and considerable training and experiment would be necessary to put the subject of air co-operation with the infantry, during an attack, on a sound basis.

Furse was quite seriously wounded by shrapnel and had a skull injury. He recovered, but was later killed, although the circumstances of his death are not clear.

Lieutenant Colonel Edmund William Furse, Chevalier of the Legion of Honor, MID (commanding 88th Brigade Royal Field Artillery) was killed in action at age 41 on the 19 May 1918, and is buried at Dormans French National Cemetery. His younger brother was killed in 1914.

His Medal index card from *Ancestry* notes that from 18th November 1914 he was to be "GSO 3rd Grade, Royal Flying Corps Headquarters" and in October 1915 he was elected as a member of The Royal Aero Club of the United Kingdom. An earlier edition of the Royal Aero Club's magazine had noted (on 17th September 1915) that Major Furse had obtained his aviator's certificate (number 1706) in a Maurice Farman Biplane at the British Flying School at Le Crotoy, France on the 8 September that year.

It appears that Collis was attached to 16 Squadron officially from the 16 June as a Sergeant pilot. 16 Squadron had formed in February 1915 at St Omer, to carry out a mixture of offensive patrolling and reconnaissance, but also did a lot of work on using wireless to report troop movements. Lieutenant Elliott was also attached to 16 Squadron at the same time.

This was an odd one as Collis recorded in his logbook arriving in France and flying at St Omer initially, but then seems to take up quite a lot of operational flying. At first, I could not place him with a squadron. 16 Squadron moved from St Omer to La Gorgue on 6 March, 1915 under the command of Major Felton Vesey Holt CMG, DSO.

16 Squadron at that time was under the command of Air Vice Marshal Felton Vesey Holt, CMG, DSO. He had started his squadron flying with 4 Squadron in April 1913 and was later posted to 16 Squadron in 1914, becoming its commanding officer. As Major and CO of 25 Squadron, he flew the F.E.2b. He remained in the RFC and on into the RAF becoming a Brigadier General in the newly established service. During the inter-war years, Holt remained in the Air Force, serving in several staff appointments before becoming Air Officer Commanding Fighting Area. Holt was killed in a flying accident not long after taking up his final appointment.

Air Vice Marshal Felton Vesey Holt CMG, DSO, RAF, the photo taken from his entry for his RAeC certificate number 312 gained on 1 October 1912. *(Royal Aero Club Trust)*

Having carried out an inspection of RAF Tangmere, he had taken off in his D.H..60M Moth, K1838, with his PA flying another Moth, when a formation of Siskins, seeing his PA's aircraft, decided to pay him an aerial salute. Unfortunately, the formation dived directly at the AVM, whom they could not see, and the 43 Squadron Siskin J8893, piloted by Sergeant Charles George on the left of the formation, caught the wing of his Moth. Going into a spin, the pilot recovered, and the AVM took to his parachute, but they were too low, the AVM being killed when he hit the ground without a fully deployed parachute. The pilot, Flight Lieutenant Henry Michael Moody, died when the aircraft hit the ground. The inquest brought a verdict of accidental death.

Collis did his first trip from this airfield on the 16 May 1915. According to his records at the National Archives, he only served with 6 Squadron operationally, but a search of the *Air History* website revealed he was attached to 16 Squadron, and so was his observer Lieutenant Elliott, so it made sense that the with the operational flying he was recording in his log book he was with a squadron. What confused the issue is that his service records at Kew do not show this, but the *Air History* website shows a record of him being attached for June and July so it's worth checking a number of sources when trying to piece together movements.

For most of June 1915, Collis and Furse are out on artillery observation duty with the occasional engine trouble for good measure, one of the observations Collis comments:

...heavies very successful

A clear indication that the guns were spot on.

On 16 June, Collis is flying with Sergeant William Whiddon Hart, RFC entry number 457, on a wireless test. Whiddon was previously a soldier in the 3 Battalion Coldstream Guards who had transferred to the RFC in October 1912 and was later posted to France in August 1914. He went on to serve with both 16 and 4 Squadrons and remained in the RAF post WW1 and on into WW2, gaining an MBE and retiring as a Wing Commander, last being noted in the Air Force List in 1962.

Collis and Furse's last flight together occurred on the 21 June 1915, having flown more than 20 hours over the front as a pair. No doubt Collis picked up a lot of useful 'gen' from this seasoned observer. On the 21 June, Collis flies his longest sortie to date over the front with Furse, a flight of 3 hours 45 minutes duration.

From the 4 July until the 14 July, Collis flies again with Lieutenant Gordon Elliot, on mainly artillery observation. Lieutenant Gordon Roy Elliot was on attachment from the 3rd (Prince of Wales) Dragoon Guards and was to be an observer from 16 June 1915. He later qualified as a pilot gaining certificate number 1872 on 11 October 1915. Lieutenant Gordon Roy Elliot had been attached to 16 Squadron with Collis since its formation at St Omer. He later served on 22, 33, 46, 57, 62 and 207 Squadrons, flying a vast variety of types, including the S.E.5a, and becoming a squadron commander in his own right. He was promoted to the rank of Major and was appointed squadron commander, but by the end of the war he was forced to resign his commission due to ill health.

On one sortie, Collis takes with him Air Mechanic John Henry Dollittle, service number 2761, on a trip from Choque to Merville aerodrome in heavy rain! Dollittle is an interesting character and worthy of further mention!

Dollittle's service records are available at the National Archives as a digital download, and his RAC Certificate and photo at *Ancestry.co.uk*. The NA ref is Records Air 76/149/177. *(Royal Aero Club Trust)*. Due to his exceptional career and awards for bravery, a Google search will bring up a number of results for this airman and the text used comes from: https://greatwarlondon.wordpress.com/2013/04/12/john-henry-dollittle-aviation-enthusiast/.

He had enlisted in the RFC as a driver. He was certainly one to attract the attention of enemy fire and was wounded a number of times. Interestingly he was also awarded a Distinguished Conduct Medal and Military Medal for his bravery whilst serving with the RFC – an unusual combination.

Dollittle joined the Royal Flying Corps in January 1915. His brother George also joined up, enlisting in the Somerset Light Infantry in November 1914 and arriving in France in September 1915. 2nd Class Air Mechanic 2761 J. H. Dollittle was sent to France much more quickly than his brother, presumably because he already knew his way around an aeroplane. He arrived on 24 January 1915.

On 10 March, an RFC aeroplane was forced to land close to the front line. Dollittle and three other 2nd Class Air Mechanics were sent out under Corporal S.C. Griggs to undertake repairs. They worked through the night under heavy shellfire and got the aeroplane back into working order, so that it was able to take off again in the morning. For their hard work, the five RFC men were awarded the Distinguished Conduct Medal.

Credit to this great website and an extract within from a family member who wrote about John, *https://greatwarlondon.wordpress.com/2013/04/12/john-henry-dollittle-aviation-enthusiast/.*

Dollittle continued to serve in the RFC/RAF for the rest of the war, becoming a Corporal in 1915 and a Sergeant in May 1916. In 1917, he was serving with 46 Squadron, a fighter squadron based at Sutton's Farm in Essex as part of the defence of London from Zeppelin and bomber aeroplane attacks. His brother George was severely wounded in the chest during the battle of the Somme in 1916 but, apparently against expectations, survived his wounds.

On Christmas Day, 1917, John Dollittle (now serving with 46 Squadron in France) was again rewarded for his bravery. His Flight Commander announced that he and two other men were being awarded the Military Medal. Like his DCM two and a half years earlier, John's award was earned through his work salvaging aeroplanes under fire.

Until late 1917, John appears to have remained unscathed, despite his bravery under fire (something to which ground crews were not often subjected). In November 1917, though, he was sent back to the UK suffering the effects of gas.

In August 1918 he suffered a much more severe injury. At that point, aeroplanes had to be started by hand by swinging the propeller. Undertaking this routine duty on a Sopwith Camel on 5 August 1918, Dollittle was struck on the arm by the propeller. Within ten days he was back in England, and after treatment that went on for years, his arm was amputated in 1920.

What an exceptional career and story!

Collis completed his time at La Gorgue with 59 hours of operational flying before being posted to No. 6 Squadron.

6 Squadron, Abeele

When researching squadrons, you will often find that there are good websites or publications that cover the squadron's role, its aircrew and groundcrew and history. This information is from Steve Buster Johnson website *https://www.stevebusterjohnson.com/*

No. 6 Squadron appeared in the Royal Flying Corps (Military Wing) Orders for 30 January 1914:

Captain J.H.W. Becke will carry out the formation of No. 6 Squadron commencing tomorrow. No. 6 Squadron will be based at South Farnborough.

The squadron deployed to France on 6 October, with the aircraft being flown via Dover to Bruges where they all arrived safely. When the first battle of Ypres commenced on 19 October, the squadron was based at Poperinghe.

No. 6 Squadron's principal roles was co-operation with the Army. The earliest recorded act of co-operation occurred on 20 October 1914 near Lille, when the pilot of a 6 Squadron aircraft, having observed shells fall on a German battery, dropped a message to gunners of the 87th Royal Field Artillery. The message, believed to have been dropped in a bottle, read:

"You hit them. We must go home. No Petrol"

During the winter months of 1915 when no effective operations were possible, the squadrons of the Royal Flying Corps were engaged in developing the various means of air co-operation which the lessons of 1914 had taught them. In the domain of wireless telegraphy great progress was made. Experience had shown that the best method of signalling from an aeroplane was by wireless, and by the beginning of 1915 most of the initial difficulties had been overcome. Reconnaissance duties also extended to the aerial photography of trench systems, and for the March offensive on Neuve Chapelle, the whole trench system of the enemy had been carefully photographed from the air.

On the 10 March, a notable operation was carried out by Captain Strange of 6 Squadron. His aircraft had been fitted with rudimentary bomb racks of Captain Strange's own design, under the wings that were operated by pulling a rope in the cockpit. They flew to Courtrai and Menin respectively where they dropped bombs on railway junctions of strategical importance to the German reinforcement of Neuve Chapelle.

The summer of 1915 saw a marked development of aerial fighting. Reconnaissance machines had to fight hard to gain information. In the spring of 1915, they rarely flew above 6000 feet, but by the summer they were forced up to 12,000 feet. Further, they were soon compelled to work in pairs, one doing the work the other acting as escort. On the 25 July, Captain Hawker of 6 Squadron was on patrol in a Bristol scout. After engaging two enemy aircraft, he successfully shot down a third. As a recognition of his determined attacks on enemy machines culminating in the above-mentioned exploit, Captain Hawker was awarded the Victoria Cross, the second to be gained by the Royal Flying Corps and the first for success in air combat.

No. 6 Squadron's work continued with observation for the artillery engaged on counter battery work, registration and corps reconnaissance. The great value of the work of 6 Squadron had a unique recognition. The squadron was specifically mentioned in Sir Douglas Haig's first Despatch dated 19 May 1916.

Collis is found in the 6 Squadron Officers' Register as a 2nd Lieutenant, with an address of 66 Belvedere Road, Burnley, Lancs. He notes in his log book that he was transferred to No. 6 Squadron and promoted in the field, to 2nd Lieutenant as of 29 June 1915. At that time the squadron were based at Abeele aerodrome. He would have been rubbing shoulders with the likes of Captain Lanoe Hawker who went on to win the Victoria

Of interest is the fact that Collis's commission from the rank of Sergeant took place in the field and is mentioned in the book *Royal Flying Corps (Military Wing) Casualties and Honours during the war 1914-17*. He was commissioned to 2nd Lieutenant under "Honours Conferred by His Majesty the King, for services in connection with the war". No doubt his work with 16 Squadron and his flying with Major Furse had something to do with this.

Abeele Aerodrome, a photo taken in 1915 when 6 Squadron were in residence, a view that Collis would have been very familiar with. *(TMAM)*

Cross and DSO, also Captain Louis Arbon Strange, DSO, OBE, MC, DFC and Bar.

So, what was happening at the front at the time that Reginald Collis arrived? Letters written home by Lanoe Hawker give some idea of the aerial war for the squadron, although Lanoe Hawker was flying offensive patrols in his Bristol Scout in between reconnaissance duties. This letter dated the 23 June 1915, a few weeks prior to Collis arriving, gives some idea – and it's clear that there was quite a bit of action:

I had bad luck with my new beast yesterday returning from a patrol. I left the ground and chased two who had the cheek to come right over us! But they turned and made for home at once and I never got near them, so I did a little patrol their side. One Hun must have got a 'norful' fright if nothing worse, climbing peacefully well his side; I dived at him from the front and opened fire at about 50 yards! But we passed so quickly that aiming was difficult and time so short. He vanished in a spiral nosedive, but then they all do that when frightened and I could not see if he flattened out safely or not. Another of them saw a couple of shells burst just close to me and shut off his engine and went down at once when about 2 miles away. Time being up I started home, and then saw a third on our side again. I thought I had him beautifully, but just as I approached my petrol gave out! And I had to make a forced landing. He saw me being Archied and turned back home at once, but I could have intercepted him if I had my engine. Hopeless country to land in just S of the floods and I took a barbed wire fence and turned upside down! Total damage to me – Breeches torn and slight bruise on right leg. Machine hors de combat but damage not vital.

Hawker VC RFC ACE: The Life of Major Lanoe Hawker VC DSO, 1890–1916,
Tyrrel M. Hawker MC

Lanoe's letter of 6 August tells us more about the enemy aircraft and tactics:

They have huge machines now, and they do give us trouble- they are certainly more upish these days and have been actually known to attack us more than once! I think they made a determined attempt to obtain the mastery, and in consequence we have had several casualties, but we've come up to scratch all right. We hear they have 'wind up' and my last letter shows they haven't had it all their own way. On the 31st I got beautifully over one, but my gun jammed and though I frightened him I didn't get him.

Last Monday in an F.E on our way home we came up behind a couple ranging. Left No. 1 diving steeply after 140 rounds, attacked and chased No. 2 well home till we got too low over their lines... and later attacked a third, who made off into the clouds. We since heard No. 1 landed behind their trenches (Confirmed by the King's Own Regiment) so evidently, we at least did his engine – unluckily there was a strong wind in his favour. They are all armed with machine guns now of course and don't hesitate to use them. Further South I believe they are giving the French an awful lot of trouble.

Hawker VC RFC ACE: The Life of Major Lanoe Hawker VC DSO, 1890–1916,
Tyrrel M. Hawker MC

It was against this backdrop that Collis was to join the squadron. He also experienced combat and encounters with enemy aircraft on numerous occasions, suffering damage from return fire.

His first patrol with the squadron occurred on the 4 August 1915. Collis took off in B.E.2c number 1718 for a reconnaissance over the lines at 6,000 feet, with one Cuthbert Julian Orde in the observer position. Orde had joined up as a 2nd Lieutenant in the Army Service Corps on 15 August 1914, transferring to the RFC and becoming a Flying Officer observer and being posted to 6 Squadron. He was a Lieutenant when he qualified as a pilot in a Maurice Farman biplane on 10 May 1916 at which time he was he was promoted to Flying Officer. Later on, in

A photo of B.E.2c Serial 1748 at Abeele Aerodrome as flown by Captain Louis Arbon Strange *(copyright Steve Buster Johnson)*

Cuthbert Orde in his RFC Observer uniform. He too had joined the Squadron in July of 1915. *(Tangmere Military Aviation Museum)*

1917, he was promoted to flight commander. He was given the rank of temporary Major on 16 August 1918 but he relinquished his commission on 15 January 1919 on grounds of ill health and retained the rank of Captain. Orde always identified as an artist and after WW1 had a studio in Paris. He is most well-known for the series of portraits of fighter pilots of the Battle of Britain, but many are not aware of his role in WW1.

On their first patrol in the air, sadly after 40 minutes Collis was forced to land in a bean field due to:

...lack of pressure in the tank, no petrol, needle of pressure gauge was stuck.

Reading 6 gallons (when actually empty) he managed a good landing despite the crops:

...beans up to height of the bottom of the main planes, machine ok.

12 August sees the pair off on a patrol in machine number 1718 for 60 minutes at 7,000 feet over the lines, with nothing much to report, followed by an "experimental photography" trip of 1.5 hours on Saturday the 14 August.

On Tuesday 17 August 1915, the pair carried out three sorties in machine number 1718. The first two sorties were unsuccessful due to being hampered by bad weather, with low cloud and poor visibility. Later on that day, the clouds lifted and the crew of 1718 climbed up to 9,500 feet to carry out their reconnaissance sortie for a duration of 1 hour 45 minutes. The route they took is recorded in the log book as:

Hollebeke, Zandvoorde, Houthem, Warneton, Comines, Wervick, Menin, Courtrai, Mouscron, Tourcoing, Roubaix, North of Lille.

All was going well until near the end of the patrol, when the engine began misfiring and Collis could not maintain height. Rather than get too low and close to the enemy lines, as well as the ever-present risk of engine failure, Collis decided enough was enough and headed home to Abeele airfield, North of Lille. His only comment was:

Archie exceedingly busy, but no hits.

'C' flight of the squadron at this time were employed on ranging the guns by wireless, and all of the reconnaissance flights were carried out by 'A' and 'B' flights, which were being commanded by Lanoe Hawker. So, Collis must have been in 'A' or 'B' at the time, based on his log entries. One of Lanoe's last combats was on the 11 August, when he was in action with six enemy aircraft in total. He had summarised his combats over the previous few weeks as:

In 12 flights out of 15 reconnaissance patrols, German machines were encountered. In 12 fights that resulted, 15 hostile machines were engaged of which five were brought down and nine put to flight.

Captain Lanoe Hawker was awarded the VC for his efforts, and was posted back to England in August. Collis was to meet him once again in January 1916 on 24 Squadron at Hounslow.

On the 22 August 1915, Collis and Orde were flying together and got airborne from Abeele in good weather for another reconnaissance flight. The duty was to take photographs in B.E.2c number 1718. They headed to the Lille area at 8,000 feet and managed to take eleven photographs northeast of Lille. Orde also made use of his artistic skills by making sketches of trenches, gun pits and the local defences. This time they were in luck and there were no problems encountered during the 1 hour 50-minute sortie. Collis recorded in his log, "very successful". This was the last time that Collis and Orde were to fly together.

Lieutenant Eynon George Arthur Bowen was the next observer that Collis was to fly with. He had arrived on secondment from the Royal Garrison Artillery on the 4 August 1915, but sadly he was not to survive the war.

Son of the late Eynon George Rice Bowen, of Troedyraur, Cardiganshire, and of Georgina Catherine

An F.E.2a of 6 Squadron, the first of its type to go into active service with the RFC and the first lost in combat. It was with 6 Squadron for about the same period as Collis was with the squadron. *(Steve Buster Johnson)*

Bowen, of Harcombes, Hambrook, Bristol. Scholar of Sherborne School. R.M.A. Woolwich Cadet. Eynon George Arthur Bowen originally joined the Royal Garrison Artillery and transferred to the Royal Flying Corps as an observer. He was posted to No. 6 Squadron where he flew with Reginald Collis. He later qualified as a pilot, flying with both 35 and 22 Squadrons on the Western Front. He was sadly to lose his life in September 1916, flying an F.E.2b when engaged by Hauptman Oswald Boelcke of Ja 2. He went down in flames and became a PoW but later succumbed to his wounds.

Collis first flew with Bowen on the 25 August 1915 in BB 2 1706 for a two-hour sortie over Ypres and Hooge at 8,000 feet. Fortunately the weather was good for artillery observation, but they encountered heavy anti-aircraft fire and the machine was damaged. Collis was to record:

Machine badly damaged by "Archie", new top left hand main plane fitted, right hand main plane damaged, hole through fuselage one foot from the pilot.

Not an uncommon entry for the time, "Archie" was fierce along the lines and between artillery shells and anti-aircraft fire, the air could 'boil' or be extremely turbulent making any artillery observation or ranging work both hazardous and very uncomfortable, often with aircraft returning with a few extra pounds of shrapnel or missing certain parts! On 26 August they set off in aircraft number 1680 for a 1-hour artillery observation flight, same as the previous day and in the same vicinity. This time they happened across an enemy machine and they engaged:

...fight with enemy machine, drove down from observing over Ypres.

Sadly, the success of the engagement cannot be confirmed. They tried a further flight, but they had to return due to "wireless trouble", however, later in the afternoon a change of aircraft and observer brought success.

This time Collis was in aircraft number 1713 with a Lieutenant Parker, heading for an area northeast of Ypres at an altitude of 8,000 feet. Collis recorded:

Zonnebeek district, artillery observation. While observing, saw enemy machine ranging. Attacked him and drove him down. He appeared to land about Menin. About an hour later, we were attacked by a fast enemy scout, he engaged us from the rear, flying slightly higher and from the direction of our lines, diving down on top of us, this was at 8000 feet over Hooge. The enemy fired 50 to 60 rounds at us before I could get into a favourable position to reply, smashing the wireless set completely. I turned sharp underneath him, forcing him to fly over and past me and by the time he had turned we had turned also, with our gun to bear on him. We flew towards each other, each firing at the other, at about 50 yards range, we repeated this manoeuvre three times when the enemy had apparently had enough or had exhausted his ammunition, for he dived for his own lines, nevertheless, completing what he had come up to do i.e., stop us ranging, as we had to return owing to the wireless being out of action.

The following day this RFC Communiqué was sent:

2nd Lt. Collis and Lt. G.A Parker on a B.E.2c, when flying over Ypres, attacked and chased an Aviatik which, upon being fired at, dived and appeared to land. A second machine, a German Scout, attacked them near Hooge and, hitting the wireless set, succeeded in breaking it. After a good deal of ammunition had been expended the German Machine flew away towards Roulers.

The observer that day was George Parker, who was originally with the Northants Regiment and arrived on secondment to 6 Squadron on the 21 April 1915. He also flew as observer to Captain L. A. Strange and was in combat in a B.E.2 on the 24 May 1915 near Zonnebeke with an Aviatik. He also flew with 13, 8 and 60 Squadrons, but was later killed in action.

He was promoted to the rank of Captain and awarded a DSO and MC. The citation for his DSO reads:

Citation 17 January 1917
Capt. George Alec Parker, M.C., North'n R. and RFC. For conspicuous gallantry in action. He attacked hostile aeroplanes on three occasions during the same flight, killing an enemy observer. On another occasion he drove off three enemy machines, pursuing one of them down to 750 feet three miles behind the enemy's lines.

He was sadly to be killed in action whilst flying a Nieuport 17 with 60 Squadron on the 27 November 1916. The victory was claimed by none other than ace pilot Lieutenant Werner Voss of Jasta 2.

Collis Flew with Parker again on the 30 August, in machine number 2031, again around the Ypres, Hooge area at an altitude of 8,000 feet. Yet again they were attacked:

Artillery observation, we were attacked by Aviatik, drove him down over the enemy's lines. He dived steeply, at the same time signalled for assistance from his anti-aircraft batteries by means of white star light, could not see him land owing to mist but when last seen was still diving practically vertical. About 35 minutes later we attacked another German machine, he didn't show fight but promptly dived.

A three-hour, action packed sortie! Interestingly the RFC communiqué for 28 August records the following action which is thought to be the action recorded as the 30th, so maybe Collis submitted his information a few days later, catching up with his sorties and getting the date wrong:

2nd Lt Collis and Lt. Parker when on a B.E.2c carrying out artillery registration near Hooge, saw a hostile machine coming to attack them from the direction of Roulers. Turning to meet it, they fired fifty rounds and then manoeuvred round again to get into a more favourable position. Having expended most of their ammunition, a wireless message was sent for a Scout to come out, but on firing the last few rounds, the German machine dived steeply. The observer then signalled to say that the German had gone. The Registration was continued, and was interrupted twice by hostile machines which, however, when approached showed no fight.

The following day Parker and Collis departed in machine number 2031 for a sortie over Hooge to carry out artillery observation at 7,000 feet. They were up for 1 hour 35 minutes before they suffered engine trouble and were forced to make an early return to Abeele aerodrome.

Collis got a few days rest before his next sortie on 4 September in aircraft 1714 with Lieutenant Morgan. They headed from Abeele to an area over the Trenches northeast of Ypres to carry out artillery observation, but the weather was set to confound them once again. With low cloud and poor visibility, they only managed to see two of the three targets. After crawling around in the murk for 1.35 Hours they returned for a well-earned cup of char!

His observer, Lieutenant Ralph Carleton Morgan, was seconded to 6 Squadron from the Canadian Artillery on the 4 August 1915, and he was to later fly on 12 Wing, RFC. After he ceased flying, he became an instructor at No. 5 School of Military Aeronautics. No. 5 School had been formed on 1 August 1917, from a cadre of officers and men from No. 2 School of Military Aeronautics, based at Christchurch College, Oxford, moving to Denham on 8 September 1917.

On the 5 September Collis flew with another observer, one Lieutenant Thomas, but they had to return from the planned sortie after just 35 minutes due to more engine trouble.

Here is the obituary from the Journal of the Royal Aero Club dated the 2 Aug 1917, which was also in *The Times* and the *Epsom Advertiser* dated 3 August 1917:

Lieutenant MAURICE WOTTON THOMAS, RFA and RFC, reported missing on August 5th last year, now officially presumed killed on that date, aged 21, was the younger son of Mr. And Mrs. Alick Thomas, of Worcester Park, Surrey. He was educated at Parkside, Ewell, and in September 1907 passed into Osborne College, where, as the result of an epidemic he contracted an illness, which necessitated his being invalided, as unfit, from the Royal Navy in 1910. After 18 months of convalescence, he was allowed to continue his education with Mr. Sellar, of St. Andrew's, and Mr. Tinniswood, of Camberley, whence he passed into Woolwich in 1912, and, passing out in September 1914, received his commission in the RFA

In May 1915, he went to the front as A.D.C. to a General, commanding an R.A. division, and was promoted Lieutenant in July, when it was decided that he was specially qualified by his training for the RFC, to which he was seconded as Observer. After many thrilling experiences he was wounded in an accident in December 1915 and came home. He made a speedy recovery and was sent to a reserve squadron and gained his wings. In May 1916, he was again sent out, doing much valuable work over the enemy positions. He was a fine revolver shot and a member of the shooting eight when at Woolwich.

His elder brother, Captain Alec Vaughan Thomas, E. Surrey Regiment, attached 2nd Hampshire Regiment, was killed in Gallipoli, August 6th, 1915, aged 22.

Maurice was awarded the 1915 Star, British War medal and the Victory medal. This information can be found on the Medal Index Cards at the National Archives, as well as on other genealogy websites.

Lieutenant Maurice Wotton Thomas of the Royal Field Artillery was attached as an Observer on probation to 6 Squadron on the 12 July, 1915. He was injured in an accident later on that year, after an engine failure on take-off caused the aircraft to hit trees. He trained as a pilot, gaining certificate number 2671 in April, 1916, and was posted to 4 Squadron, where sadly on the 5 August 1916, whilst flying in machine number 2649 (a B.E.2d), he was shot down by anti-aircraft fire northwest of Bapaume and killed. *(Royal Aero Club Trust)*

On 6 September 1915, Collis heads out on a solo sortie, flying machine No. 1718, to attack troops around Lichtervelde. He records in his log book:

Dropped 100lb bomb on concentration of troops, also two boxes of darts. Archie did good shooting, met enemy machine headed for him (bluff, no guns on board) but he dived without showing fight.

By darts, Collis is referring to steel *flechettes* which, on first look, are no different from ordinary darts. In reality they are of a highly aerodynamic and deadly design, constructed of steel with a sharp pointy end and fins on the other end to facilitate a smooth flight. These *flechettes* were dropped from the sky onto enemy troop concentrations as early as 1914 by the French. The RFC and Germans soon followed suit.

On 7 September, Collis is back over Hooge with a new observer, this time with Lieutenant J.E.P. Howey, for a sortie lasting just over 1 hour. The weather this time played ball judging by the comments by Collis as weather nited as "Normal". He records the sortie as:

Photography, enemies reserve line of trenches around Hooge, J7-J25, Archie troublesome, we attacked Aviatik, gun jammed had to retire.

Lieutenant J. E. P. Howey – John Edwards Presgrave Howey – was an interesting character and an eccentric millionaire to boot! Howey by all accounts was not a regular army officer and he had joined the Imperial Yeomanry and then served in the Dorsetshire Yeomanry and the Bedfordshire Yeomanry before transferring to the Royal Flying Corps. Howey's family had owned considerable property in Melbourne, Australia and on his father's death (John Edward Werge Howey) the family fortune passed to his elder brother and on his death in 1901, passed to John. John appears to have only served on 6 Squadron and was also shot down whilst flying with them.

> Howey's brief service records can be found at the National Archives under AIR 76/241/39 but they only confirm that he moved to the RFC from the Bedfordshire Yeomanry and the date he went missing.

On 11 November 1915, Kelway-Bamber failed to return from a patrol in his F.E.2. His observer, Lieutenant J. E. P Howey, was taken prisoner. As an English millionaire he was paraded around as a bit of a trophy. His health suffered and he was repatriated to the care of his wife, Gladys, as long as he didn't return to the UK for the duration. This led to him living in Switzerland until 1918.

Letter sent by Howey 28 January 1916:

I was captured at Courtrai at 10 a.m. on the 11th of this month (November), and I met two German Officers there who knew several English people that I knew, and they were most awfully kind to me. They gave me a very good dinner of champagne, oysters, etc, and I was treated like an honoured guest. I then came by train the next day to Mainz, where I was confined in a room by myself for two days. I have now been moved into a general room with eight other English Officers where we sleep and eat. We are treated very well and play hockey and tennis in the prison yard. Poor B——, I was so sorry he was killed, he was such a nice boy and only 19. I had a fight with two German aeroplanes and then a shell burst very close to us. I heard a large piece whizz past my head, and then the aeroplane started to come down headfirst spinning all the time. We must have dropped about 5,000 feet in about 20 seconds. I looked round at once and saw poor B—— with a terrible wound in his head and quite dead. I then realised that the only chance of saving my life was to step into his seat and sit on his lap where I could reach the controls. I managed to get the machine out of that terrible death plunge, switched off the engine and made a good landing on terra firma. I shall never forget it as long as I live. The shock was so great that I could hardly remember a single thing in my former life for two days. Now I am getting better, and my mind is practically normal again. We were 10,000 feet up when B—— was killed, and luckily it was this tremendous height that gave me time to think and act. I met one of the pilots of the

Sourced from *https://www.stneotsmuseum.org.uk/a-letter-home/lieutenant-j-e-p-howey-royal-flying-corps-great-staughton/*

> *German machines which attacked us. He could speak English quite well and shook hands after a most thrilling fight. I brought down his machine with my machine gun and he had to land quite close to where I landed. He had a bullet in his radiator and petrol tank, but neither he nor his observer were touched.*

After the war, Howey set up the Romney Hythe and Dymchurch railway! The RH&DR was the culmination of the dreams of Captain J. E. P. Howey – a racing driver, millionaire landowner, former Army Officer and miniature railway aficionado and Count Louis Zborowski – eminently well-known racing driver of his day (famous for building and racing the *Chitty Bang Bang* Mercedes).

On the 9 September 1915, Collis is airborne with Lieutenant Bowen again for a spot of patrol duty around Hooge area at 8,000 feet but nothing to report. A sortie of 1 hour 50 minutes.

The 10th sees a change in observer, back with Lieutenant Maurice Wotton Thomas who was an officer of the Royal Field Artillery. This time they set off on a patrol duty over the lines at 8,000 feet for just over two hours.

Later the same day, Collis and Lieutenant Morgan got airborne in 2674 for a 3-hour sortie doing artillery observation around the Hooge District at 8,000 feet. The following day Collis was paired with Lieutenant E. W. Leggat, MC, of the Wiltshire Regiment.

Squadron Leader Leggatt, M.C. was born in 1892 and joined No. 6 Squadron in France as an observer in May 1915. In August 1915 he had several brushes with the enemy, the most interesting being on the 25th:

> *Lieutenant Cooper-King and Lieutenant Leggatt set out in an F.E. armed with two machine guns. When over Polygon Wood at a height of 11,400 feet the Pilot saw a hostile Biplane over Hooge and dived to 7,200 feet, getting within 100 yards of the enemy. Both machine guns were fired simultaneously, and the German aeroplane was seen to dive, turning in the direction of Menin... apparently hit.*

After a brief spell with 21 Squadron as a pilot at the end of the year, he joined 10 Squadron on 2 February 1916, and on the 29th fought his MC action:

> *2nd Lieutenant E.W. Leggatt and 2nd Lieutenant T.S. Howe (BE.2c, 10 Squadron) while on patrol sighted an Albatross north of the Bethune-La Bassee Canal, flying towards the lines at about 6,500 feet. 2nd Lieutenant Leggatt again chased and slowly overhauled the German, opening fire at 150 yards. The hostile machine dived, followed by the BE.2c still firing. The German's propeller was now seen to slow down, and his machine dived more steeply. The B.E.2c continued diving after him until within 1,000 feet of the ground and keeping up a fire all the time. The hostile machine was seen to land and turn upside down about two miles south of Merville. The Pilot of the hostile machine was wounded in the leg and he and his observer were taken prisoner. Tracer bullets are reported by Lieutenant Leggatt to have been of considerable assistance.*

Promoted to Captain in May 1916, Leggatt joined No. 2 Squadron that same month, and in August was posted missing when it was reported:

> *Captain Leggatt... went up in the Bethune-La Bassee neighbourhood on 9.8.16. Machine fell in German lines, but still under control.*

The following information is obtained from the publication *Medal Circular*, Issue No.13, August 1999, produced by Spinks, which advertised the sale of medals awarded to Squadron Leader Edward Wilmer Leggatt, MC (Royal Australian Air Force, late Wiltshire Regiment, Royal Flying Corps and Royal Air Force). Often medal auctions, back catalogues and websites can give useful information on service records, careers, awards and medals.

Leggatt was next heard of as a Prisoner of War in Germany, and after about a year was moved to *Offizier Gefanenen Lager*, Holzminden, which "flung open its hospitable gates to its English guests" in September 1917.

Shortly afterwards, the elderly Commandant was removed and the reign of the villainous Hauptman Karl Niemeyer (notorious for the victimization of Leefe-Robinson, VC) commenced. A few escape attempts were made in the early days, but not one officer succeeded in crossing the bounding river Weser and Niemeyer "became blatantly cocksure that a successful escape from his camp was a total impossibility." While numerous British officers persisted with high profile antics, a small group of dedicated types took up the challenge in earnest and the "Holzminden Tunnelling Company Ltd." swung into action.

For nearly all of the nine months it took to complete the 60-yard-long tunnel, its existence was only known to the 13 men of "the working party". And during these long nine months, its presence was never suspected, despite the fact that the Camp Commandant prided himself on being the most alert jailer in the Fatherland. In the second week of July 1918, the working party – in the grip of escape fever – let a chosen few, Leggatt among them, into the secret of the impending escape, though kept them in ignorance of the tunnel's whereabouts. After one false start everything was set for the night of 23 July, but at the eleventh hour an officer making an "over the wire" attempt nearly jeopardized the whole venture.

Edward Wilmer Leggatt, M.C. Born in Dehra Dun, India on the 12 February, 1892 and educated at Rugby, he joined the Wiltshire Regiment and transferred to the RFC, serving in France as Observer during 1915. He later trained as a pilot and flew on the Western Front in 1916 but became a prisoner of war, though he was not captive for long! He survived the war and later resigned from the RAF After the war he served in the RAAF *(Royal Aero Club Trust)*

His plan, however, was discovered by the "Tunnelling Company" and he was unceremoniously ejected from their building and told to make excuses to any German he might meet in returning to his own quarters. Meanwhile, Leggatt and his fellow travellers, Tullis and Purves, put the finishing touches to their escape kits. The final "Go" was whispered by "Munshi" Gray of the Indian Army in Hindustani, 24 hours later. By this time the atmosphere in the camp was electric as almost everyone except the Germans, of course, realized a major break-out was in the offing. At 11:30 p.m., the working party entered the tunnel and made good their departure. After an agreed clear hour, six officers who had assisted with vital tasks were given the green light, together with the S.B.O. (Senior British Officer) and two other favoured officers. Then between tense pauses, a handful of officers at the top of the "waiting list" were despatched until one, overladen with hiking gear, became jammed and a halt was called.

A total of 29 officers went through the Holzminden tunnel, and of these, ten reached England safely. The first man home was the portly S.B.O., Colonel Rathbone, who travelled by train all the way to Holland on a forged passport. Three members of the "worming" party (do they mean *working* party?) were the next over the frontier, where they were joined in a Quarantine Camp at Rijks by Leggatt, Tullis and Purves on 19 August. A remarkable achievement indeed, when it is considered that "Out of some 8,000 officers in Germany, a mere handful – between 40 and 50 – succeeded in regaining their freedom."

Back in England, Leggatt served at the CFS in 1919 but was soon demobbed. In 1923 he was granted a

Short Service Commission as a Flying Officer in the RAAF, which lasted until 1926. He returned to the Service in 1940, became Commanding Officer of 3 and 7 E.F.T.S and finally relinquished his commission at the end of the Second World War.

Leggatt's Military Cross was awarded in 1916 and is recorded in the *London Gazette* dated 30 March 1916:

For conspicuous gallantry and skill when on patrol. 2nd Lieutenants Leggatt and Howe attacked a hostile machine and drove it down. They then climbed again and attacked another. Under heavy fire from this machine, by a combination of good flying and shooting, they hit the enemy pilot and damaged his engine, forcing him to descend within our lines. The enemy were made prisoners.

He was awarded a Bar to his Military Cross for his escape from captivity, as mentioned in the *London Gazette* dated 16 December 1919:

In recognition of gallantry in escaping from captivity whilst Prisoners of War.

Leggatt and Collis first flew together on the 11 September 1915 and this was a long reconnaissance in machine number 2674 at 14,000 feet. Collis records the route in his log book:

Crossed the lines at 10,000 feet, Reports on Lille, Tourcign, Tournai, Courtrai, Quesnoy, Roubaix, Menin, Bousbecque, Wervicq, Comines, Landygorde, Hollebeke, Archie asleep, no shells at all, no hostile machines encountered, trip very successful.

12 September sees Collis off on a pratice reconnaissance with Lieutenant Howey and on the 13th he carries out an artillery observation with Lieutenant Morgan. This sortie around the Hooge district is at 10,000 feet for a sortie length of 2 hours and 35 minutes. These sorties must have been quite draining, not just the constant vibration and noise of the engine, but also the cold and exposure. Crews were no doubt running on adrenalin to get the sortie completed.

The following day the pair were not so lucky, with aircraft number 1713 having a big-end seizure at 4,000 feet on the way out to a patrol. Luckily they had enough height to glide back and make a safe landing. 19 September they had better luck completing a patrol duty at 7,000 feet in good weather.

The same day, Collis headeds off with an observer he had not flown with before, Lieutenant Terence Donough O'Brien, on a photography sortie around Hooge at 8,000 feet in machine 1718. The flight was succesful.

O'Brien was to be killed whilst with 6 Squadron after returning from a reconnaissance flight on 3 March 1916 at Abeele, near Poperinghe, Flanders. The accident occurred while landing on returning in aircraft 4324, a B.E.2c. The pilot on that flight was Lieutenant Pier-

Terence Donough O'Brien originally joined the 16th Lancers. He was the only son of Brigadier General Edmund Donough John O'Brien, CB (14th King's Hussars) and Florence Harriet O'Brien, daughter of Henry Wheeler of Worcester Park House, Worcester Park, Surrey, and grandson of Lieutenant Colonel Sir John Terence Nicholls O'Brien KCMG, Governor of Heligoland and Newfoundland. He was born on February 28th, 1896. He joined his regiment (16th Lancers) in Flanders on October 23rd, 1914, and was present in the action of February 21st, 1915, in the trenches at Zillebeke, when the 16th Lancers lost a lot of men. He then transferred to the Royal Flying Corps in September 1915, as an Observer and flew with Reg Collis. *(TMAM)*

pont, who was lucky to be only slightly injured. Sadly, Terence O'Brien was not so lucky. He is buried at Lysenthoek Military Cemetery near Poperinghe.

The next few days, Reginald Collis is mainly engaged on Artillery observation and "registering flashes", but the weather closes in on the 27th and the reconnaissance scheduled for that day with Lieutenant Howey is cut short due to low clouds at 1,000 feet. On face value, this seems to be enough height to fly, but at that altitude, it would have put them at significant risk from small arms and machine gun fire.

On the morning of 27 September, Lieutenant R. Collis went out in his B.E.2c, serial 2674, on an artillery observation mission to drop a 100 lb bomb on the railway line between Courtrai and Turcoing. Again, the weather was not to play ball and he ran into some low cloud. After flying east for some time, he became unsure of his position. As light was fading fast, he decided to go home so he turned and flew due west.

After a time, he came down through the clouds and found himself over Bailleul. He dropped his bomb with safety clip attached in a ploughed field, so that it did not explode. He then flew round Bailleul until No. 1 Squadron put out flares, with the help of which he landed successfully. Collis recorded in his log book:

Bomb 100lbs, to be dropped on railway line between Courtrai + Tourcoing, very cloudy, clouds at 1,000 feet, flew due E, came down through clouds somewhere between Menin + Courtrai on River Lys, climbed over clouds again, flew easterly direction, came down through clouds again, but failed to get my bearings again, by this time it was practically dark. Decided to come home, flew due West, I came down through clouds over Bailleul, dropped bomb with safety clip attached, in ploughed field, flew round Bailleul until No. 1 Squadron put out flares + Landed, 6-20 p.m. quite dark.

This must have been quite testing for Collis as he had had no instrument flight training, and trying to maintain level flight using a compass and altimeter, and relying on his senses alone, which were flawed. All pilots know that they should rely on what their instruments tell them to remain in controlled flight. Sadly Collis had few to go on.

Pilots of the Royal Flying Corps were at the mercy of a westerly wind constantly trying to blow them over enemy lines, which meant flying longer back into wind to get home when fuel reserves were dwindling! The wind could be stronger at altitude, and if you could not see the ground, getting lost was almost a dead cert!

You could end up being blown much further off course than expected, leading to disorientation when coming out of cloud to look for a familiar feature. If you had not set your altimeter correctly, descending through cloud could result in hitting trees, hills, buildings etc. with almost a certainty of death. The odds were certainly stacked against the poor pilot trapped above cloud. With night falling fast, making a landing as soon as possible was the only option. Louis Strange had a similar close call as he explains in his book *Recollection of an Airman*, which illustrates the perils of getting lost above cloud or in poor visibility:

...we had been flying for about three quarters of an hour, and as we could not get a single glimpse of the ground, I decided to go through the clouds and look where we were. I was rather worried at finding my altimeter show only 2,000 feet when we touched the top of the cloud layer, but we sank down into the clouds and continued to drop until we registered 500 feet. Then I suddenly remembered that in the hustle and bustle of our start I had forgotten to set the instrument at zero. I knew that the aerodrome at St Omer was 400 feet above sea level. Matters were decidedly serious, because I had anticipated no difficulty in gliding down through the clouds, but now it looked as if there was fog on the ground. Resolving to go up again and have another look round, I opened up the engine, but as I was not practiced in cloud flying, I found great difficulty in keeping the compass from swinging and the pitot tube from jumping about. I Throttled down again and decided to descend until I could see the ground, meanwhile keeping

my machine in as stalled an attitude as possible in case I had to make a sudden jump. The state of my nerves can be imagined when I saw my altimeter record 300 feet, 200 feet, 100 feet, 50 feet and then 0, and still found myself in the midst of a dense, impenetrable fog. It was so thick that I could not see the interplane struts, and as the darkness of evening was coming on apace I found it difficult to read my instruments correctly. Down I went, cursing myself for a fool for not having set the altimeter correctly before starting my flight. If I had done so, however, I should have probably been even more frightened than I was, for suddenly the mist cleared and I nearly stalled the machine in my astonishment when I looked down and saw below me a nasty, green, choppy sea!

I was thankful I had not hit the ground, but nowhere near solving the next problem – where was I? I speculated whether I had been drifted off the Belgian coast by a high, southerly wind when flying between 6,000 and 7,000 feet, or overshot the aerodrome at St Omer and been carried back over the Channel by an East wind.

Recollection of an Airman, Louis Arbon Strange

October dawned with Collis and Lieutenant Leggatt back together, carrying out photographic sorties and "experimental" photography, plus some experimental bomb dropping. Things seemed to be going well – until they suffered an engine failure at 500 feet, right after take-off – just to keep them on their toes!

Fortunately, they were able to make it safely back to the airfield.

There is grave danger in turning back at that height. Low level engine failure often offers no choice but to land ahead. The higher you go, the greater the temptation to turn back towards the airfield and try and make it to a nice piece of field upon which you know you won't turn over on your back. The pitfall here, is that without power the aircraft is descending rapidly, and pulling tight angles of bank to reach where you think you want to go only increases the stalling speed of the machine. The all too common result is a stall – and spinning into the ground with a bone-crunching crash! Close monitoring of the airspeed is essential to ensure survival.

Collis made his first flight in an F.E 2b on 7 October, in machine number 5643. A 15-minute sortie to gain experience on the type. The following days saw further photographic sorties with Lieutenant Leggett on the 10 and 11th. Collis then seems to get a few days break from flying before being paired with another observer, Lieutenant J. E. Griffin, on 23 October. Griffin was attached to the RFC from the 4th Ox and Bucks Light Infantry for duty as an observer. He was posted in on 23 October, so this flight must have been one of Griffin's first flights with 6 Squadron. He later left the squadron on the 6 November 1915 – shortly after an engagement with the enemy while flying with Collis!

Lieutenant Morgan is back in the observer position on 24 October for a 'ranging' sortie, but the fog gets the better of them after 70 minutes of flying around at 4,000 feet and they land just behind the Belgium trenches. They fly the aircraft back to the airfield later the same day. The 26th sees the pair off on a 2-hour artillery observation in machine 5644 at 8,000 feet, but there was nothing to report.

Lieutenant J. E. Griffin was to get a baptism of fire with Collis on 30 October when they took off on a two-hour patrol flight in machine number 5643. The patrol was at 9,000 feet when they were attacked and Collis notes in his log:

Patrol duty attacked by enemy twin tractor + Aviatik. We dived down onto the Aviatik, which was 1,000 feet lower than we were, firing half a drum when our gun jammed, at the same time I observed twin sticks, diving down on us, firing about 500 yards range. We headed for him which caused him to turn to his right and away from us, at the same time I turned sharply to the left + put my nose down + flew over our own lines, failing to repair gun, came home.

The next day he is flying with Flight Sergeant Thomas Bosworth, who was an early entry into the RFC as number 309. Enlisted in 1905 as a gunner in the Royal Garrison Artillery, he later transferred to the RFC with a posting to 3 Squadron on 6 August 1912. He was posted to 6 Squadron on 6 October, 1914 and is promoted to Flight Sergeant (fitter-turner) on 1 May 1915. Bosworth was to survive the war and went on to serve in the RAF until 1933. A classic example of a tradesman being taken aloft for use as a gunner/observer.

The sortie was for a duration of 1 hour 35 minutes over the Hooge district, taking photographs. Whilst engaged, they witnessed a dog fight above them between an FE and a Focker. This combat was more than likely 2nd Lieutenant C. H. Kelway-Bamber and his observer, Lieutenant H. J. Payn from 6 Squadron, who were trying to chase a Fokker away from another B.E.2c No. 2031. It was 12:45 and they were flying an F.E.2a over Zillebeke when the combat occurred.

Collis makes the following entry:

Photography, while taking photos, witnessed duel between FE + Focker, saw Focker turn over on his back and nosedive, obviously out of control + crash about Zillebeke.

The rest of November sees Collis testing a machine, and a photography test with an entry for tuition on 8 November. The tuition flight was with Lieutenant Edward Henry Paul Cave of the Army Service Corps, who learned to fly and gained certificate number 1537 at Farnborough.

As Cave was taken on strength with 6 Squadron on 6 November 1915, it's highly likely that Collis was showing him the local area and landmarks; a bit of familiarisation and local procedures.

On the 17 January 1916, a month after Collis had departed the squadron, Cave was still flying with 6 Squadron. He was flying a Bristol Scout over the trenches when he was wounded by a rifle bullet from a German Aviatik, and had to force land during a patrol of the Ypres salient. He remained in the RFC and became an instructor but was later posted overseas. It is not clear who was teaching whom on the trip, but it was only 20 minutes in duration.

2nd Lt E. H. P. Cave. *(Royal Aero Club Trust)*

11 November 1915 sees Collis undertake his last successful photography sortie, with Flight Sergeant. Bosworth. The flight was over Hooge and lasted 1 hour 20 minutes with nothing much to report. The next day he was to undertake his final flight in France, although this was cut short due to poor weather and low cloud.

Something clearly happened to Collis, as he stops flying completely, with no further entries for the year with 6 Squadron. His service records indicate he is admitted to Wimereux 14th General Hospital.

14th General Hospital where Collis ended his days operationally. The reason for his hospitalisation is sadly speculation but likely to have been fatigue or stress. *(TMAM)*

Reginald is sent home on 29 November 1915 and his records show that he is then discharged to duty on the 18 December 1915.

It is not clear what struck him down, but he does have a brief spell back flying in 1916. However, by September of that year he is taken ill again, this time whilst flying again and his career as a pilot ends.

On reflection of his time on 16 and 6 Squadron, he certainly had his fair share of adventures, damage, engine failures, combat and getting lost above cloud, to keep any aviator on the edge of his nerves, and I wonder if he was mentally exhausted, however there is nothing in his records to indicate this.

Having flown for 30 years and experienced engine failures, partial engine failures, getting lost, smoke in the cockpit, carbon monoxide exposure and bad weather – I can say that one's nerves do get tested. I was never a natural pilot and was always scared of heights, yet I could happily fly an open cockpit biplane at 7,000 feet.

Add in anti-aircraft fire, enemy aircraft, exposure and hypoxia through lack of oxygen – and I can imagine that some found the whole experience stressful, and were probably only driven on by the thought of not letting the side down. Failure was not an option! Or maybe the thought of fighting in the trenches was worse! What was the lesser of the two evils? They had all witnessed the suffering and fate of the poor bloody infantry!

His flying summary for the end of 1915 shows Collis has completed 187 Hours 58 minutes in the air, of which 128 hours were operating over the Western Front with 6 and 16 Squadrons.

He had survived without getting shot down and with no injury!

His log book indicates that as of January 1916 he is considered fit to fly, and is posted to the newly forming 24 Squadron at Hounslow, now commanded by Major Lanoe Hawker VC, DSO, his old flight commander.

24 Squadron, January 1916

24 Squadron was being formed under the command of Major Lanoe Hawker VC, DSO at Hounslow in September 1915, and was equipped with the Vickers F.B.5. There would have been many 6 Squadron pilots interested in joining the new Squadron, as they looked up to Lanoe Hawker and would want to fly under his leadership.

Vickers F.B.5 as flown by Reginald Collis at Hounslow in September 1915. *(Tangmere Military Aviation Museum)*

Collis takes his first flight at Hounslow with Captain R. E. A. W. Hughes-Chamberlain.

Hughes-Chamberlain had gained his certificate at Brooklands on 20 February 1915 on a Maurice Farman Biplane, gaining certificate number 1094. He initially flew with 11 Squadron before joining 24.

Hughes-Chamberlain joined 24 Squadron in December 1915, while it was still working up at Hounslow Heath. He led "B" Flight until 16 August 1916, when he was forced to land at Baisieux in D.H.2 5929 after being wounded in combat. Hughes-Chamberlain would go on to be a flight commander attached to No. 4 Squadron, Australian Flying Corps, from November 1917 to 18 February 1918. He would also earn the Air Force Cross.

On the 14 January, Reginald Collis took his first flight in a Vickers F.B.5, which went well, but on the 19th of the month he was tasked with taking an F.B.5 from Farnborough to Hounslow. Without any warning, the engine let go a valve and the stem of the valve buried itself in the rear spar of the aircraft. Any structural damage to these early aircraft was potentially life-threatening, any crack or damage to a spar could, if mishandled, or if it occurred in rough weather, lead to a failure and a one-way ticket down without a parachute.

After only a few flights with the squadron, for some reason that is not recorded, Reginald was then posted to No. 1 Reserve School at Farnborough, with the purpose of ferrying aircraft to France. Mostly taking machines from Farnborough to St Omer, his posting occurred with effect from the 21 January 1916. No explanation can be found as to why this was, and we will never know. One thing is for certain, ferrying aircraft across the channel was by no means risk free!

On 27 January, Collis made his first trip in an Airco D.H.2, and all went well. A few days later, on the 5 February, he made his first ferry flight across the channel to St Omer from Farnborough. In the remarks column of his log book he states:

From time leaving ground Farnborough to Landing at St Omer = 88 minutes.

These trips were fraught with danger, although lifebelts were available in case of an unplanned dip in the sea due to mechanical failure, weather over the channel could be very changeable and, in winter, any forced landing and exposure to the cold water could be fatal. Fog and low cloud could also make any journey across stressful.

Airco D.H.2 as flown by Reg Collis on his first ferry flight across the channel in 1916. *(Mark Hillier Collection)*

The next few weeks saw Reg ease into his new job, and he ferried an F.E.2c to St Omer, with a short stop at Folkestone due to a snowstorm, but after a false start he made it satisfactorily to France. He was not so lucky on the 14 March when he was tasked with flying a B.E.2. Again, the weather was against him and Collis ran into fog three miles from Calais. His entry in his log book on the 14th states:

Ran into fog, four attempts at landing, finished up in a dyke, smashing undercarriage + prop (Coquelles 3 miles from Calais).

Flying was at its most basic. If you couldn't see the ground and couldn't get over the fog or cloud to a clearer location, you would either gingerly feel your way down through the murk, anxiously looking at your altimeter and hoping that you had remembered to set it correctly, but run the risk of hitting terrain or buildings.

The only saving grace with having fuel left was that you could put the power on and go up for another try.

If you searched about for too long you could run out of fuel and have to glide down through the murk anyway, keeping your speed as slow as possible without stalling, in the hope of reducing the force of the inevitable impact with the ground. All very nerve-wracking to say the least! All that said, Collis would have known that the chances of escaping without damage were pretty slim!

On the 19th of the month, Collis set off once again from Farnborough with his observer Flight Lieutenant Emery – in good weather at last – in an F.E.2b heading for St Omer. They crossed the Channel at 8,000 feet (the higher the better in case of an engine cough!). However, en-route they spotted an enemy aircraft. Collis records:

Shot raider down in mid-channel 8 miles NE of Deal (see report).

The extract from the Daily RFC communiqués for the combat over the channel on the 19 March 1916 reads:

2nd Lt. R.Collis and Flt. Lt. Emery (F.E.), while flying over the channel at 8,000 feet, saw a hostile machine being shelled over Dover. They gave chase but could not get within range. Presently they saw another machine making for Deal. They flew up channel and met this hostile machine at 4,000 feet. Planing down with engine throttled back until within 150 yards, the observer opened fire immediately behind. The hostile machine did not return fire and made no attempt to manoeuvre out of range. After a drum had been expended the enemy was observed to plunge down towards the sea with a steep right-hand bank and with irregular puffs of smoke coming from the engine. Lt. Collis now experienced some difficulty with his engine and during the time that he was changing over the petrol to the service tank, he lost sight of the hostile machine, which was last seen at 1,500 feet diving steeply.

Reginald kept a newspaper clipping of the time in his log book that reads:

Second Raider Reported Brought Down.

The press association says information has come to hand that during the raid on Dover last Sunday, a second German seaplane was brought down by a pilot who was engaged in taking a new machine to the front. If the report is true, the pilot evidently then continued on his way to France. (No date or details on clipping).

Extract of information from *http://www.iancastlezeppelin.co.uk/19-mar-1916/4590506241*. These events often made the local newspapers as they were over the channel or home territory. A Google search will often produce results, as in this case. Ian Castle has a brilliant site that focuses on WW1 air raids on Britain which gives an excellent insight as to events on the day and help illustrate the one-line entry in the logbook.

19 March 1916

On 19 March, SFA1 planned another of their Sunday daytime seaplane raids on Kent coastal towns. This one proved the most lethal to date. The attacking force comprised six aircraft: four Friedrichshafen FF33s, a Hansa-Brandenburg NW and a Gotha Ursinus WD, although it is possible only four or five of the aircraft dropped bombs on land.

At 1:57 p.m. the first of three incoming aircraft appeared over the Admiralty Harbour, Dover, at a height estimated between 5,000 and 6,000 ft, from which one aircraft dropped the first bomb, followed by three that fell on Northfall Meadow, close to Dover Castle. One smashed into a hut housing men of the 5th Battalion, Royal Fusiliers. The blast killed two soldiers, Frank Roseberry and Walter Venables, mortally wounded two more and injured another 11. Turning near the castle, the raider then dropped bombs in Castle Street, followed by bombs along Folkestone Road. One destroyed a tram track and the blast sent a cyclist, Miss Edith Stoker, aged 23, smashing into a shop doorway at 131 Folkestone Road. She died in hospital. That bomb also killed seven-year-old Francis Hall who was on his way to Sunday School. Two more bombs fell in Folkestone Road, causing much damage. The raider then turned back towards the harbour, dropping a bomb in Northampton Street that killed 47-year-old Mrs Jane James and injured two other women. The final bombs landed at the harbour; one in Wellington Dock and two in the commercial harbour.

By now a second aircraft had begun bombing Dover. At Eastbrook Place, a bomb hit an orphanage run by the Sisters of St. Vincent. The children rushed down to the basement when the first bombs exploded in the town, and the only casualty was one of the Sisters, falling roof tiles cutting her arms. Further bombs fell close together in Church Street, King Street, Russell Street and at the corner of Woolcomber and Trevanion Streets, causing varying amounts of damage. The final bombs fell near the harbour; two in East Cliff and three in the sea.

Two of the three raiders then followed a course north towards Deal, where one aircraft dropped nine bombs on a line from south to north. The first three fell close together, one wrecking Woodbine Cottage in Victoria Road. The other two fell in a neighbouring garden destroying two chicken houses, killing a chicken and blasting tree branches up into the telephone lines. The rest dropped between High Street and West Street, one exploding by the graveyard wall of St. George's Church, but there were no personal injuries.

Then, at about 2:12 p.m., two other German aircraft approaching Ramsgate from different directions, began dropping 14 high-explosive bombs over the town. One bomb landed in the road close to St. Luke's church with devastating effect. The bomb landed on a car causing the petrol tank to explode. It tossed the driver, 49-year-old Harry Divers, up into the air, killing him instantly. The blast also killed four children and mortally wounded another, aged between four and 12, on their way to Sunday School. A 23-year-old woman, Mrs G.M. Bishop, died two days later from her injuries. Another bomb landed on a stable close by from where the blast smashed dozens of windows.

Nine bombs fell around the gasworks without causing significant damage and one fell on the home and shop of a hairdresser, Mr. T. Desormeaux, in Chatham Street. Mr. Desormeaux and his family rushed outside when they heard explosions, just as a bomb hit the roof, causing considerable damage to the upper storey, but they all escaped injury. Another bomb struck the Chatham House School, used as a hospital for wounded soldiers by the Canadian Army. The bomb smashed through the roof but there were no injuries. Another bomb damaged the premises of W. P. Blackburn, Upholsterer and Undertaker, at 71-73 King Street

The final bomb of the raid was a single H.E. bomb dropped on Margate. It landed on a house at 29-30 Fort Crescent, Cliftonville, used as headquarters of the 9th Provisional Cyclist Company, damaging the roof and dislodging chimneys.

The attack had stirred up a response, and the RFC and RNAS had 26 aircraft in the air from Dover, Eastchurch, Grain and Westgate. There was also a F.E.2b being ferried to France that successfully joined in the engagement. Two of the German raiders, an FF 33 and the Hansa-Brandenburg, were brought down on the return flight, both forced to land at sea. One made repairs and managed to fly back to Zeebrugge while the other had to

F.E.2b aircraft that Collis ferried to St Omer across the Channel on 1st March 1916, as well as claiming an enemy aircraft over the channel on 19 March, flying with Flight Lieutenant Emery. *(Collis Collection)*

An F33, a type that attacked Dover on the 19 March 1916. *(TMAM)*

A Hansa Brandenburg was the other type used during the raid on Dover, and it was this type Collis had combat with. Collis had actually seen the raiders heading into Deal, so he waited and positioned his FE to catch the raiders on the return home. The aircraft he attacked was crewed by Lieutenant Fredrich Christiansen (who later became Germany's leading naval airman) and his Observer Oberleutnant-Zur-See Von Tschirschky. This chap was the Zeebrugge Commanding Officer. The aircraft was hit in the radiator and put down but was later repaired and flown back to base. This was the first time that the F.E.2b was used in any Home Defence action. *(TMAM)*

be recovered by German vessels and towed into port.

Up to this point in his career, Reginald Collis had been in combat on eight occasions, but this sortie was the only one that we can definitely say he had a confirmed success. The others were mostly inconclusive or never confirmed. He either recorded that he drove the aircraft down or lost sight after the contact. At least his success was some consolation for the casualties that day, 14 killed, 26 injured and nearly £4,000 of damage to property.

Further ferry flights followed in a B.E 2C and a Martinsyde – although the latter flight ended in a bad landing on rough ground with both a broken axle and propeller! On the 10 April he flew a D.H. Scout across the Channel which took him 1 hour 30 minutes, but this ended in a forced landing, a quarter of a mile from St Omer. Luckily, the machine was repairable, and was able to be flown out of the field. Collis was on a run of bad luck and on 13 April, he was tasked with testing a B.E.2C which was apparently armoured. His entry in the log book says it all – and he was very lucky to survive! The weather, he records as impossible with gusty winds:

> *...testing machine, machine took charge owing to strong wind (gale lasted 4 days) machine crashed into gasometer RAF Building [Royal Aircraft Factory] damage, machine total wreck, myself broken collar bone + concussion, also suffering from shock.*

Collis spent the next three months recuperating from his prang and does not fly again until the 10 August, when he got back in the cockpit and completed a 20-minute solo in the B.E.2c.

A Martinsyde, the type that Collis first flew on 31 March 1916, ending up in a landing accident with broken axle and propeller. He was lucky it did not turn over as this example did. *(Mark Hillier Collection)*

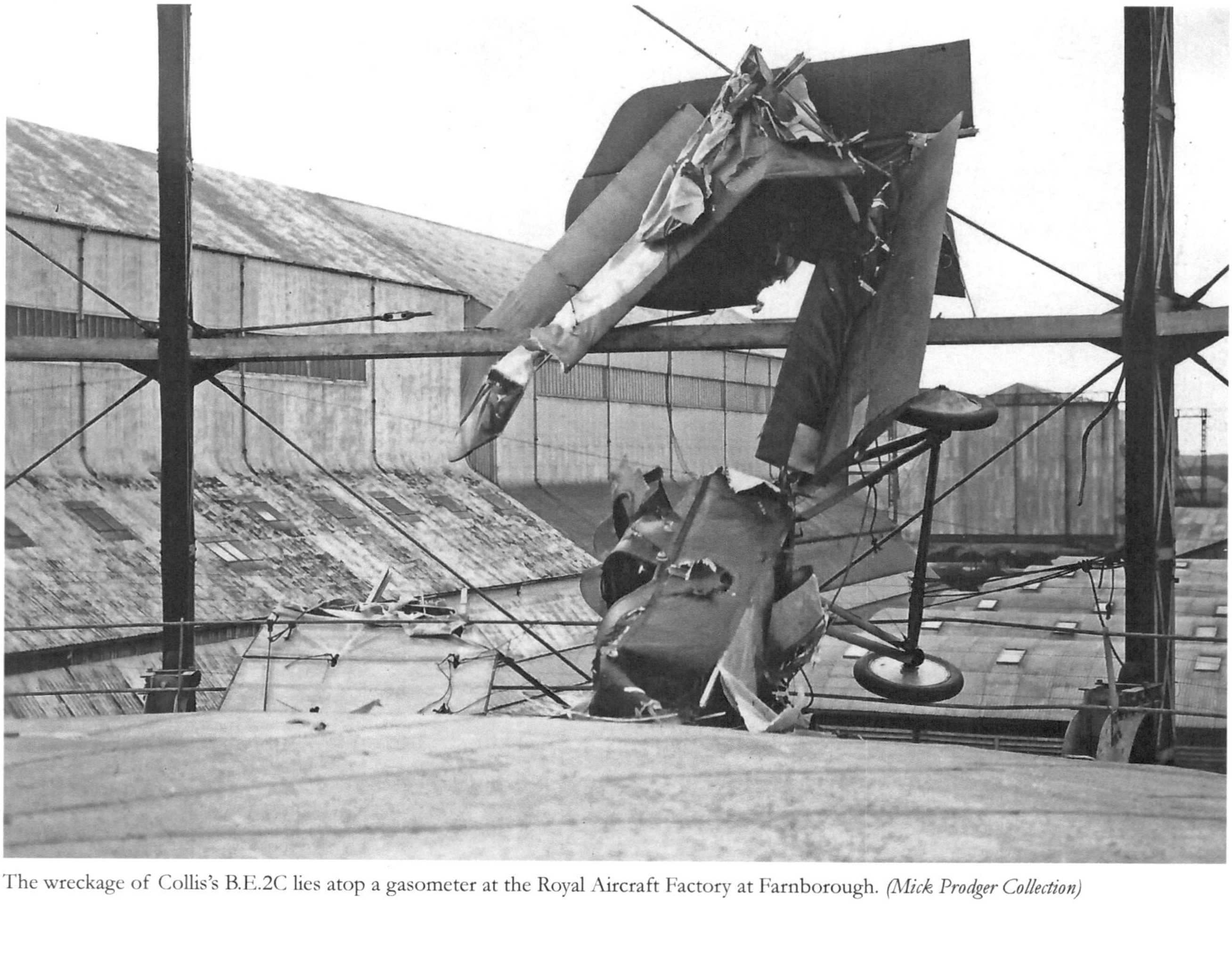

The wreckage of Collis's B.E.2C lies atop a gasometer at the Royal Aircraft Factory at Farnborough. *(Mick Prodger Collection)*

Collis being made comfortable following his collision with the gasometer in which he suffered concussion and broke his collarbone. The remains of the upper wing of his B.E.2C can be seen wrapped around the stanchion of the gasometer. *(Mick Prodger Collection)*

On 20 August 1916, Collis was sent to Glasgow to pick up an F.E.2d, machine number 4926, and bring it ultimately to Farnborough. His passenger is recorded as a Mr. Henry Bremner. The pair set off south towards Oxford, but again the weather was not great, as Collis recorded:

...cloudy and rain weather unsuitable for flying.

But things go from bad to worse when they encounter engine trouble near to Oxford after flying for five hours:

...forced landing, valve tappet broken, also petrol ran out, mechanic filled main petrol tank with 9 gallons and oil, disconnected same and ran on front tank, made temporary tappet from a bolt, engine ok.

On 23 August, Collis flew the aircraft out of a field and back to Farnborough, where the machine was fully repaired and tested. The following day he was back in the air with a ferry trip to St Omer in an F.E.2c. Although there were strong winds and low cloud, he made the trip with no further issues.

His next ferry trip occurred in a Martynside Scout, 160 hp, on 2 September and in bad weather again, noted as "cloudy and rain" in the comments section of his log. This flight also ended with mechanical failure near Guilford – a broken camshaft this time – and once again put him in a field!

Collis found himself posted to 41 Squadron at Gosport on the South Coast. 41 Squadron had re-formed on 14 July 1916 with a nucleus of men from 27 Reserve Squadron and was equipped with the Vickers F.B.5 Gun Bus and Airco D.H.2 Scout.

These were replaced in early September 1916 with the Royal Aircraft Factory F.E.8, and it was these aircraft which the squadron took on their deployment to France on 15 October 1916. Collis took his first flight in a Royal Aircraft Factory F.E.8 on the 19 September, from Gosport, and managed to return to earth safely but his luck was not to last. The following day he took off in an Avro 80 HP for a trip to Cowes on the Isle of Wight, but the Gremlins were at play again, and this time a broken inlet valve meant another forced landing.

For Collis, the end of his flying career came on the 30 September, when he took off in an F.E.8 for Farnborough in cloudy weather, clearly a bumpy day according to his notes. After a stressful few months of engine trouble and forced landings, it must have taken a toll on his nerves. He simply wrote:

Landed owing to feeling ill in the air.

After 230 odd hours of flying, the log book entries cease.

It is not clear what the issues were. Maybe fatigue, stress, or a medical issue played a part. In his service record it only states "...fit for ground duties not general duties" with no further details entered. Was he at the end of his nerves? Had the combat experience mixed with the precarity of flying pushed him too far? He was not alone, that's for sure. Many others turned to drink to help solve their problems. Some could not continue and were returned home.

When you search the National Archives, from time to time you find letters that give some idea of the stresses and strains on aircrew. A letter from a medical officer to the 5th Brigade Royal Flying Corps Headquarters, written in 1917 about one of the pilots of 21 Squadron, gives us an insight into some of the problems faced and is somewhat typical of the signs and symptoms of combat stress.

This officer joined the infantry at the beginning of the war, being then 16 years old. He served in the ranks until June 1915 when he got a commission in the North Staffords. He came to France on September 1st, 1916, was through the Beaumont Hamel fighting and in the trenches until December 1916. He then came straight to the RFC as observer on probation, without any course at home, and was with No. 46 Squadron until March 1917, when he came to No. 21 Squadron.

Until five weeks ago he felt perfectly fit, but then began to be troubled with Insomnia and bad dreams, and to feel generally tired and depressed. He went on leave a week later which he spent quietly at home and was beginning to feel better, but on his return the old symptoms recurred. He further began to avoid the society of his brother officers, and to make a practice of keeping a flask of whisky by his bed, with the hope of getting some rest at night by that means. On one occasion he apparently disgraced himself when under the influence of alcohol.

On examination he presents the signs of one who is generally tired and depressed. His pulse rate is increased above the normal at rest, his heart being in that "imitable" state frequently found amongst those suffering from the strain of service. As regards the connection of alcohol with his case, I consider that it is not responsible for his condition but may have aggravated his previous symptoms. I think that he needs a good rest under supervision but should completely recover with care. At present he is temporarily unfit for service.

Letter found at the National Archives, ref: TNA AIR/11547/204/78/6

Reginald Collis had certainly done his bit, and even after returning home he continued to fly, even though it is clear that the stress was perhaps getting to him; shorter flights, regular reports of bad weather, frequent prangs. I am only speculating that maybe he was at the end of his tether. He was a brave man, as they all were, and to endure the days of early wartime aviation was clearly not only exhilarating, but at times also a white-knuckle ride. Aviation being more an art form than a science at this point in time.

Despite his short flying career, he certainly had some varied experiences from being an instructor near the outset of RFC flying, through to combat experience in France, meeting lots of interesting characters and personalities on the way. At the time of his operational flying on the Western Front with 16 and 6 Squadron, life expectancy was not too bad: around the 200 to 300 hours flying time – compared to 1917 when life expectancy could be measured in weeks and maybe 50 - 60 hours flying time. In some ways Collis was lucky to be at the front in 1915 as aerial combat was somewhat in its infancy, with ground fire and mechanical or structural failure presenting a bigger danger than enemy aircraft. Collis himself suffered damage to his aircraft from the dreaded "Archie" but was never shot down.

Had Collis gone back to operations in 1916 or 1917 it is highly likely he would have ended up a casualty.

The time that Collis was on 6 Squadron, from July to November 1915, the squadron suffered four casualties killed in action, two prisoners of war and one wounded in action.

Figures by Steve Buster Johnson *Over the Front.*

Although the log book is not exceptional in terms of Collis's length of service on operations, and he was not an ace, he most definitely saw combat and miraculously survived many perils. His log book offers a fascinating insight into the early days of military aviation and a reminder of the dangers he and ohers faced on an almost daily basis. It provides an interesting snapshot of history, illustrated by researching service records and photographs, as well as contemporary first-hand accounts and texts. Flying log books tell a story and are great for historians and collectors to get their teeth into – and this one is no exception. Once you start adding faces to names and maps to places, the story just comes alive.

While future generations may not have the pleasure of knowing the brave men who took to the air and pioneered military aviation in WWI, the log book of Reginald Collis offers a unique insight into one such man, and brings his story to life.

The Log Book

PILOT'S LOG BOOK.

Name

Place

Date from _____ to _____

Rank_____

Name_____Reginald Collis_____

Place_____

SUMMARY OF FLIGHTS.

Week Ending	Time in Air.		Machine.	Engine.	Aerodrome.		Cross Country.		Reconnaissance.		Passenger.
	Hrs.	Mins.			Hrs.	Mins.	Hrs.	Mins.	Hrs.	Mins.	

Height.	Distance.	Weather.	Remarks.

<u>TYPES OF MACHINES FLOWN</u>   Total

MF.	= 1912 pattern	70 H.P. Renault	
"	= 1913 "	" "	
"	= 1914 "	" "	
Short-horn		80 H.P. Renault	
HF	= 1914	80 H.P. Gnome	
B.E. 2.a	=	70 H.P. Renault	
" 2.b	=	" "	
" 2.c.	=	" "	
" 2.c.	=	100 H.P. R.A.F	
" 2.e.	=	" "	
" 2.d.	=	" "	
" 12	=	120 H.P. R.A.F.	
F.E. 2.a	=	120 H.P. A.D.	
" 2.b.	=	160 H.P. A.D.	
F.E. 8	Scout	100 Mono	
DeH.	Scout	100 Mono	
Martinsyde	Scout (super)	120 A.D.	
"	" "	160 A.D.	
AVRO	=	50. H.P.	
"	=	80. H.P.	
Vickers Fighter	=	100 Mono.	
Curtiss JN4			
Lynx Avro		Total 21 types.	
DH4		901P	
Henry Farman			

RECORDS OF FLIGHTS.

Date.	Machine.	Passenger.	Time in Air. Hours.	Minutes.	Course.
1913.	M. Farman				
Jan: 13	403	Collis		46	Aerodrome, C.F.S.
" 14	403	do		51	do
" 17	403	do		28	do
Jan: 17	403	—		83	do
" 18	do	—	1	8	do
Jan: 22	do	—		38	do
Feb: 1st	M.F. 415	—		40	do to Netheravon
" 1	do	Lieut: Kennedy		13	Aerodrome, C.F.S.
" 11	403	—		10	do
" 12	do	—		15	do
" 22	do	—		14	do
" 25	do	—		13	do
" 26	do	—		16	do
" 27	do	—		10	do
" 29	425	—		15	do
March 6	do	—		19	do
" 7	427	A/m Gardiner		16	do
" 12	411	A/m Lloyd			
" 13	427	A/m Warren		16	do
" 13	411	A/m		23	do
" 15	411	Vitty		15	do
" 17	425			4	do
" 18	do	Lt. Vernon		10	do
" 18	411	Lt. Urquart		10	do
" 18	do	A/m Warren		10	do
Total time	—		9	43	

Height.	Distance.	Weather.	Remarks.
1,500 ft		Fair	First tuition flight with Lt. Longmore
1,000 ft		Good	Second do do
1500 ft		do	Third do do
Total time instruction 2 hrs 5 mins			
1800 ft		Good	First solo flight
1500 ft		do	Passed 1st half of Royal Aero Clubs "Brevet" landed 19 yds from centre.
1500 ft		do	Passed 2nd half for "Brevet", landed 10 yds from centre.
Graded 2nd Class Flier 5 hrs 14 mins			
1500 ft		Good	
1500 ft		do	First Passenger flight
1000 ft		do	
1500 ft		do	
1600 ft		Fair	
1000 ft		do	
900 ft		Good	
1700 ft		do	
1500 ft		do	
1700 ft		do	
1000 ft		do	
1000 ft		Fair	
700 ft		Good	
1500 ft		do	
500 ft		do	
1000 ft		do	
1000 ft		do	
800 ft		do	

RECORDS OF FLIGHTS.

Date.	Machine.	Passenger.	Time in Air. Hours.	Time in Air. Minutes.	Course.
Time carried forward			9	43	
March 18	427	A/m Surman		17	Aerodrome C.F.S.
" 18	do	Sgt. Robbins		21	do
" 25	411	——		15	do
" 25	do	——		13	do
" 25	do	——		18	do
" 26	do	——		13	do
" 27	411			18	do
" 31	do	P.O. Hogan		19	do
" 31	do	Tel. Sterling		15	do
April 8	do			5	do
" 11	429	P.O. Hogan		50	C.F.S, Andover, Lark Hill
" 11	411	——		15	Aerodrome C.F.S.
" 11	429	P.O. Hogan		50	C.F.S. to Salisbury
" 15	do	do	2	0	Salisbury & Underwood
April 22	411	——		15	Aerodrome C.F.S.
" 22	do	——		18	do
" 24	do	——		25	do
May 20	418	——		12	do
" 27	411	A/m Turner		17	do
" 27	do	——		15	do
" 27	do	——		27	do
" 28	do	Tel. Sterling		11	do
" 28	do	A/m Smith		14	do
" 28	do	P.O. Hogan		10	do
Total time			18	56	

Height.	Distance.	Weather.	Remarks.
900 ft		Good	
1,500 ft		do	
1,000 ft		do	
1,000 ft		do	
1,000 ft		do	
900 ft		do	
1000 ft		do	
1,000 ft		do	
1,000 ft		do	
500 ft		do	
2,000 ft		do	Wireless Experiments.
800 ft		do	Engine test.
2,000 ft		do	Wireless Experiments.
2,000 ft		do	do Flew to Salisbury, then to Grendon Underwood, Bucks. Landed on rough ground, machine badly strained, dismantled & returned by road.
300 ft		Good	Upavon
1,000 ft		do	
1,000 ft		do	
800 ft		Very bumpy.	
800 ft		Good	Tuition, first pupil
800 ft		do	
900 ft		do	
800 ft		do	
800 ft		do	Tuition
700 ft		do	do

RECORDS OF FLIGHTS.

Date.	Machine.	Passenger.	Time in Air. Hours.	Minutes.	Course.
Time carried forward			18	56	
May 28	411	a/m Webb		19	Aerodrome, C.F.S.
" 28	do	a/m Savill		10	do
" 28	do	Turner		11	do
" 28	do	a/m Savill		11	do
" 28	do	P.O. Hogan		19	do
" 28	do	Tel: Sterling		19	do
" 28	do	a/m Turner		19	do
" 28	do	———		6	do
		a/m Savill		34	do
" 28	do	———		10	do
" 28	do	a/m Webb		18	do
" 28	do	———		30	do
" 28	do	a/m Smith		19	do
" 28	do	a/m Savill		17	do
" 28	do	———		11	do
" 28	do	P.O. Hogan		18	do
" 28	do	Tel: Sterling		22	do
" 28	do	Lt: Wilson		11	do
June 3	do	a/m Smith		18	do
" 3	do	a/m Savill		20	do
" 3	do	P.O. Hogan		16	do
" 30	403	———		15	do
July 11	427	a/m Baker		14	do
" 14	411	a/m Dark		12	do
" 22	do	———		18	do
" 24	do	———		20	do
" 28	429	———		14	do
" 30	411	———		36	do
Total time			25	23	

Height.	Distance.	Weather.	Remarks.
500 ft		Good	Tuition
do		do	
do		do	
do		do	
do		do	
do		do	
do		do	
do		do	Engine test.
1500 ft		do	Tuition.
200 ft		do	Engine test.
500 ft		do	Tuition
1000 ft		do	"
200 ft		do	Tuition.
do		do	do
do		do	Test.
300 ft		do	Tuition
do		do	do
200 ft		do	Passenger
260 ft		do	Tuition
800 ft		do	do
500 ft		do	do
1,000 ft		do	Spiral, landing good.
1800 ft		do	do
1600 ft		Bumpy	"
900 ft		Very bumpy.	"
1000 ft		do	Engine test.
500 ft		Good	do
500 ft		Gusty	"

RECORDS OF FLIGHTS.

Date.	Machine.	Passenger.	Time in Air. Hours.	Minutes.	Course.
Time carried forward			25	23	
July 31	411	—		25	Aerodrome C.F.S.
August 1.	do	—		15	do
Sept: 15	431	—		14	do
" 17	do	—		12	do
" 20	do	—		18	do
" 20	do	—		20	do
" 25	do	—		15	do
" 26	do	—		25	do
" 29	do	—		15	do
Oct: 3	do	—		33	do
" 9	do	—		23	do
" 13	do	—		20	do
" 16	do	—		10	do
" 22	do	—		15	do
" 22	do	—	1	0	Devizes, Malbro' & back
" 24	450	—		10	Aerodrome C.F.S.
" 30	431	—		20	do
Nov: 7	450	—		10	do
" 7	427	—		10	do
" 11	do	—		27	do
" 19	do	—		45	do
" 19	450	—		12	do
" 22	do	—		15	do
" 24	do	—		15	do
" 24	427	—		17	do
Dec: 10	do	—		24	do
" 11	do	—		36	do
" 13	do	—		30	do
Total time for year 1913			35	14	

Height.	Distance.	Weather.	Remarks.
1600 ft		Gusty	
800 ft		do	Velocity increased with altitude
400 ft		Good	Hot air pipe blown off in air.
800 ft		do	
700 ft		do	
1,000 ft		do	
1,900 ft		do	
1,000 ft		do	
1,000 ft		do	
4,000 ft		do	Spiral decent, good landing.
2,000 ft		do	
1,000 ft		do	
600 ft		do	
600 ft		do	Glide from 600 ft with engine stopped.
2500 ft		do	Re-qualified for 2nd Class Flier.
700 ft		do	Engine test.
900 ft		do	
500 ft		do	
do		do	
700 ft		do	
5,200 ft		do	Spiral decent.
500 ft		do	Engine test.
do		do	
400 ft		do	
300 ft		do	
500 ft		do	
3,000 ft		do	
do		do	

RECORDS OF FLIGHTS.

Date.	Machine.	Passenger.	Time in Air. Hours.	Minutes.	Course.
Time carried forward			35	14	
1914. Jan: 21	427			25	Aerodrome, C.F.S.
" 21	do			30	do
" 22	do			25	do
" 24	do			15	do
" 26	do	A/M Smith		15	do
" 26	do	Lieut: Shepherd		55	do
" 28	do			12	do
Feb: 3	do			20	do
" 5	450			12	do
" 9	427			12	do
" 10	do			35	do
" 13	do			30	do
" 16	do			40	Andover, Tidworth & Bullford.
" 18	do			10	C.F.S. Aerodrome
" 21	do			20	do
" 23	do			24	do
" 26	do			20	do
" 26	do			26	do
March 2	do	A/M Moody		10	do
" 3	do			25	do
" 9	do			10	do
" 18	450	A/M Cummings		10	do
" 19	427	Pt Vincent		13	do
" 24	450			10	do
" 25	do			50	do
" 26	do			12	do
" 28	do	Sgt: Farren		25	do
" 28	do	A/M Fisher		15	do
Total time			45	20	

Height.	Distance.	Weather.	Remarks.
500 ft		Good	
800 ft		do	
1000 ft		do	
1000 ft		do	
1000 ft		do	
1000 ft		do	
500 ft		do	
500 ft		do	
600 ft		Bad	Gusty, good landing.
300 ft		Good	
1,000 ft		do	
800 ft		do	
5,000 ft		do	Forced landing, ignition wire broken,
1000 ft		do	engine cut out at 3000 ft, landed in
600 ft		do	ploughed field, good landing, water
500 ft		do	in jet, repaired same, engine O.K.
500 ft		do	
700 ft		do	
500 ft		do	
700 ft		Gusty	
500 ft		Good	
500 ft		do	
500 ft		do	
500 ft		do	
2000 ft		do	
500 ft		do	
5000 ft		do	
500 ft		do	

RECORDS OF FLIGHTS.

Date.	Machine.	Passenger.	Time in Air. Hours.	Minutes.	Course.
Time carried forward			45	20	
1914.					
March 31	450	C/M Stroud		15	Aerodrome C.F.S.
April 2	do	A/M Latham		15	do
" 2	do	A/M Cosgave		13	do
" 2	do	A/M Hillier		13	do
" 3	427	C/M. Pack		10	do
" 3	450	do		10	do
" 3	450	Pt Crampton		16	do
" 8	do		1	20	Upavon, Swindon & Hungerford.
" 14	450			15	Andover
" 15	do	Sgt. Turner		20	Aerodrome C.F.S.
" 20	do			57	Devizes & Scend.
" 21	do	Servant Lund		20	Aerodrome. C.F.S.
" 22	do	A/M Day		15	do
" 28	431	Mr Simond		10	do
" 26	do	Servant Moore		11	do
" 26	do	Servant Gallop		10	do
" 27	450			20	do
" 27	do			20	do
" 29	431			35	do
May 14	463			20	do
" 20	do			15	do
" 23	do			20	do
Total time			52	20	

Height.	Distance.	Weather.	Remarks.
500 ft		Good	
500 ft		do	
200 ft		do	
200 ft		do	
200 ft		do	Flew back out to 450, which had broken down
100 ft		do	owing to ignition wire disconnecting.
			repaired same & flew back back in 450.
100 ft		do	"
3000 ft		do	Forced landing at Hungerford, engine
			missing badly, water in jet No: 5 & 6.
			cylinder plugs sooted, repaired same. O.K.
			(Superior Brevêt test)
1000 ft		do	
1500 ft		do	
3,000 ft		do	
2000 ft		do	
500 ft		do	
400 ft		do	
do		do	
500 ft		do	
1,000 ft		do	
1,000 ft		do	
4,300 ft		do	(Superior Brevêt test.) Spiral decent
			engine cut out at 4,000 ft, landed 30 ft
			from mark. O.K.
2,000 ft		do	Spiral decent, good landing, broke
1,000 ft		do	strut in righthand tail boom, cause
3000 ft		do	turning too fast on ground, wheels turning
			over & tail bumping on ground.

RECORDS OF FLIGHTS.

Date.	Machine.	Passenger.	Time in Air. Hours.	Minutes.	Course.
Time carried forward			52	20	
1914.					
July 14	428	—		38	Aerodrome, C.F.S.
" 15	do	—		30	do
August 7	do	—		20	do
Dec: 1	490	—		20	do
" 31	481	—		15	do
Total time for 1913 + 1914 =			54	23	
Total time for 1913 =			35	14	
Total time for 1914 =			19	9	
1915					
January 21	725	—		35	Aerodrome C.F.S.
" 21	433	—		45	do
Total time before proceeding overseas.			55	43	
Total time			55	43	

Height.	Distance.	Weather.	Remarks.
4000 ft		Good.	Spiral decent.
3,500 ft		do	do
4000 ft		do	do
4000 ft		do	do
2000 ft		do	
2000 ft		Good	—————"—————
800 ft		do	—————"—————

RECORDS OF FLIGHTS.

Date.	Machine.	Passenger.	Hours.	Minutes.	Course.
Time carried forward			55	43	
1915 May 10th	S.H.M.F. 1857	—		15	Aerodrome, St Omer, France
" 10	do	—		25	do
" 10	do	—		25	do
" 16	do	—		30	Le Gorgue
" 16	do	—	1	15	do
" 17	5008	—		10	do
" 17	5004	Capt. Bradley	1	50	
" 20	1857	—	1	0	Le Gorgue
" 21	do	—		17	do
" 22	5004	Lt. Elliott	1	5	
" 24	1857	Lt. Elliott	1	0	
" 25	1857	—	1	15	
" 29	do	Lt Elliott		30	
" 29	do	—		35	Aerodrome, Le Gorgue.
" 30	do	—		30	do
" 30	do	—		10	do
June 1	do	Lt Elliott	1	50	
" 2	do	—		25	La Gorgue to Chocques
" 2	do	Lt Elliott	2	0	
" 4	do	Lt Bush	1	5	
" 5	do	Lt Bush	1	50	
" 7	do	Lt. Elliott	1	15	Railway △ La Basse
" 7	do	Lt Elliott	1	5	do
" 8	do	Lt. Elliott	2	5	—
" 13	do	Major Furse	1	30	
" 15	do	Major Furse	2	45	Lens
Total time			82	45	

Height.	Distance.	Weather.	Remarks.
1000 ft		Good	First flight in France, also on S.H.
do		do	
do		do	
3000 ft		do	
7,500 ft		do	Practice Reconnaissance
500 ft		do	
6,000 ft		do	Reconnaissance
6,000 ft		do	Climbing test, 6000 ft in 45 mins.
?		do	Landed owing to engine trouble.
6000 ft		do	Reconnaissance
6000 ft		do	Photography experimental, too cloudy.
6000 ft		Cloudy	Bombing expedition. 6-20 tos
2000 ft		Good	Unsuccessful owing to engine trouble.
2000 ft		do	Engine test.
2000 ft		do	do
2000 ft		do	do O.K.
8,400 ft		do	Ranging for Artillery
2000 ft		Very cloudy	
7000 ft		Good	Wireless Reconnaissance
6000 ft		do	Practice Reconnaissance
8000 ft		do	Practice Wireless Reconnaissance
6000 ft		do	Successful.
6000 ft		do	do
8000 ft		do	Practice Photography
6000 ft		do	Wireless Reconnaissance
8000 ft		do	Wireless Rec: "Archie" doing good shooting. large piece of shell imbedded in r-hand short undercarriage strut, section of lefthand top tail boom shot away, top main plane & tail plane damaged, machine 1857 disgarded.

RECORDS OF FLIGHTS.

Date.		Machine.	Passenger.	Time in Air.		Course.
				Hours.	Minutes.	
Time carried forward				82	45	
June	16	5027	Sgt Hart		55	Hazebrook
"	16	do	Major Furse	3	10	
"	17	do	Major Furse	2	35	
"	17	do	Major Furse	1	40	
"	18	do	Major Furse	1	40	
"	19	do	Major Furse		15	
"	19	do	Major Furse	2	0	
"	20	do	Major Furse	2	0	
"	21	do	Major Furse	3	45	
"	21	do	Major Furse	2	27	
"	22	do	Lt. James		16	
"	29	do	A/m Dollittle		15	Chocque to Merville Aerodrome
July	4	do	Lt. Elliott	2	10	
"	9	do	Lt. Elliott		35	
"	10	do	Lt. Elliott		18	
"	10	do	Lt. Elliott	1	40	
"	12	do	A/m Bush	1	28	
"	13	do	A/m Bush		35	
"	13	do	A/m Bush	3	30	
"	14	do	A/m Bush	1	20	Patrol duty
Total time in France as Sergeant Pilot =				59	36	
Total time				115	19	

Height.	Distance.	Weather.	Remarks.
3000 ft		Normal	Testing new wireless set.
7000 ft		do	Artillery observation, heavies, very successful
9,500 ft		do	Artillery observation
7000 ft		do	do
7000 ft		do	do
500 ft		do	Engine missing, returned.
7000 ft		do	Artillery observation
7000 ft		do	do
7000 ft		do	do
6000 ft		do	do
500 ft		do	Engine trouble.
3000 ft		Bad.	Flew through a thunder storm, raining.
6000 ft		Normal	Artillery observation
6000 ft		do	do
6000 ft		do	do
6000 ft		do	do
6000 ft		do	do
6000 ft		do	do
6000 ft		do	Artillery observation, petrol exhausted.
6000 ft		do	glided to aerodrome, good landing.

Transferred to No. 6 Squadron, promoted in the field, to 2 Lieut June 29th 1915.

RECORDS OF FLIGHTS.

Date.	Machine.	Passenger.	Time in Air. Hours.	Minutes.	Course.
Time carried forward			175	19	
July 26	B.E 2-a	—		25	Aerodrome, St Omer
" 27	B.E.2.C.	—		20	do
" 29	B.E.2.C	—		46	Aerodrome, Abeele
August 3	B.E 2.C. 1813	—		20	do
" 4	1817	Lt. Orde		40	over the lines
" 10	1718	—		45	—
" 12	do	Lt. Orde		60	
" 14	do	Lt. Orde	1	50	
" 17	do	Lt. Orde		55	
" 17	do	Lt. Orde	1	28	
" 17	do	Lt. Orde	1	45	
Total time			125	33	

Height.	Distance.	Weather.	Remarks.
1000 ft		Normal	First flight on tractor, three landings, fairly good, a little too fast.
1000 ft		do	Two landings, very good, passed out without any personel tuition, weather very gusty.
500 ft		do	First flight on B.E.2.c, landed to fast, tyre burst, wheel collapsed, machine slewed round sharply & turned up on nose. damage :- new wheel & prop: otherwise O.K.
1000 ft		do	Three landings, first two perfect, third fair.
6,000 ft		do	Forced landing in bean field reason :- lack of pressure in tank, no petrol, needle of pressure gauge stuck, reading 6 galls when actually empty, good landing, beans up to hight of bottom main planes, machine O.K.
5000 ft		do	Practice trip.
7000 ft		do	Patrol duty.
8000 ft		do	Photography Experimental
5,000 ft		Cloudy	Reconnaissance, unsuccessful, clouds.
7,000 ft		do	Reconnaissance, unsuccessful, clouds
9,500 ft		do	Long Reconnaissance route :- Hollebeke, Landyzgorde, Houtem, Warneton, Comines, Wervicq, Bousbecque, Menin, Courtrai, Mouscron, Tourcoing, Roubaix & N of Lille. Engine missing badly, loosing hight, came home N of Lille instead of S. of Lille. Archie exceedingly busy, but no hits.

RECORDS OF FLIGHTS.

Date.	Machine.	Passenger.	Time in Air. Hours.	Minutes.	Course.
Time carried forward			125	33.	
August 22	1718	Lt. Orde	1	50	Lille etc.
" 25	1706	Lt. Boden	2	5	Ypres, Hooge.
" 26	1680	Lt. Boden	1	0	Ypres, Hooge.
" 26	1680	Lt. Boden		15	Aerodrome
" 26	1713	Lt. Parker	1	15	Ypres N.E.
" 27	2031	Lt. Parker	1	45	Ypres N.E.
Total time			133	43	

Height.	Distance.	Weather.	Remarks.
8000 ft		Good	Eleven photographs of N.E. of Lille defences, sketches of trenches, gunpits etc very successful
8000 ft		do	Artillery observation, machine badly damaged by "Archie," new top left-hand main plane fitted, right-hand main plane damaged, hole through fusalage one foot from pilot.
8000 ft		do	Artillery Observation, fight with enemy machine, drove him down from observing over Ypres.
500 ft.		do	Landed owing to Wireless trouble.
8000 ft		do	Artillery observation
8,000 ft		do	Zonnebek district, artillery observation While observing, saw enemy machine ranging, attacked him & drove him down. He appeared to land about Menin. About an hour later, we were attacked by a fast enemies scout, he engaged us from the rear, flying slightly higher, & from the direction of our lines, diving down on top of us. This was at 8000 ft over Hooge. The enemy fired, 50 to 60 rounds at us, before I could get into a favourable position to reply, smashing the wireless set completely. I turned sharply underneath him, forcing him to fly over, & past me, & by the time he had turned, we had turned also, with our gun to bear on him. We flew towards each other, each firing

RECORDS OF FLIGHTS.

Date.	Machine.	Passenger.	Time in Air. Hours.	Time in Air. Minutes.	Course.
Time carried forward			133	43	
August 30	2031	Lt. Parker	3	0	Ypres, Hooge
" 31	2031	Lt. Parker	1	35	Hooge
September 4	1714	Lt. Morgan	1	35	N.E. Ypres
" 5	do	Lt. Thomas		35	Returned to Aerodrome, engine trouble.
" 6	1718	—	1	35	Lichtervelde
Total time			142	3	

Height.	Distance.	Weather.	Remarks.
			at the other, at about 50 yds range, we repeated this manoeuvre three times when the enemy had apparently had enough, or had exhausted his ammunition, for he dived for his own lines, never the less, completing what he had come up to do i.e. stop us ranging, as we had to return owing to wireless being out of action.
8000 ft		Normal	Artillery observation, we were attacked by Aviatik, drove him down over the enemies lines. He dived steeply, at the same time, signalled for assistance from his Anti-aircraft batteries, by means of white star lights, could not see him land, owing to mist, but when last seen, was still diving practically vertical. About 35 mins later we attacked another German machine, he didn't show fight, but promptly dived.
7000 ft		do	Artillery observation, engine trouble, returned to aerodrome.
8000 ft 3,000 ft		do	Artillery observation, very cloudy, three targets, 2 O.K.s
8000 to 12,000 ft		do	Dropped 100 ℔ bomb on concentration of troops i.e. fête at Lichtervelde (some extra turn) also two boxes of darts, Archie did good shooting. Met enemy machine headed for him (bluff, no gun on board) but he dived without shewing fight.

RECORDS OF FLIGHTS.

Date.	Machine.	Passenger.	Hours.	Minutes.	Course.
Time carried forward			142	3	
September 7	1718	Lt: Howey	1	10	Hooge
" 8	2674	Rigger		20	Aerodrome, Abeele
" 9	1718	Lt: Bowen	1	50	Hooge
" 10	2674	Lt Thomas	2	20	
" 10	2674	Lt Morgan	3	5	Hooge
" 11	2674	Lt Leggett	2	35	Long reconnaissance
" 12	1713	Lt Howey	1	25	
" 13	2674	Lt Morgan	2	35	Hooge
" 14	1713	Lt: Morgan		30	
" 19	2674	Lt: Morgan	1	0	
" 19	1718	Lt: O'Brien	1	40	Hooge
" 21	2674	———		15	Aerodrome
" 21	1740	Lt Thomas	1	20	
" 22	2674	Sgt: Nuttall		15	Aerodrome
" 24	1740	Lt Thomas		55	
" 25	2674	Lt Thomas	2	0	
" 27	2674	Lt Howey	1	0	
Total time			166	18	

Height.	Distance.	Weather.	Remarks.
6000 ft		Normal	Photography, enemies reserve line of trenches round Hooge. J7 - J25 Archie troublesome. We attacked Aviatik, gun jammed, had to retire
2000 ft		do	Test.
8000 ft		do	Artillery observation, Hooge district
8000 ft		do	Patrol duty
8000 ft		do	Hooge District 8 O.K.s
14,000 ft		do	Crossed the lines at 10,000 ft. Reports on Lille, Tourcoign, Tournai, Courtrai, Quesnoy, Roubaix, Mouscron, Menin, Bousbecque, Wervicq, Comines, Warneton, Houtine, Landygorde, Hollebeke, Archie asleep, no shells at all, no hostile machines encountered. Trip very successful.
8000 ft		do	Practice Reconnaissance
1000 ft		do	Artillery observation, Hooge District.
4000 ft		do	Big end bearing seized, returned to aerodrome.
7000 ft		do	Patrol duty
8000 ft		do	Photography. Hooge district.
500 ft		do	Engine test, missing badly.
8000 ft		do	Artillery observation.
1000 ft		do	Engine test, running well 1650 R.P.M air
8000 ft		do	Registering flashes.
7000 ft		do	Artillery observation.
6000 ft		very cloudy	Short reconnaissance, unsuccessful owing to clouds at 1,000 ft

RECORDS OF FLIGHTS.

Date.	Machine.	Passenger.	Time in Air.		Course.
			Hours.	Minutes.	
		Time carried forward	166	18	
September 27	2674		1	20	
" 28	2674	Lt Howey		20	Bailleul to Abeele
October 1	2674		1	40	
" 3	2674	Lt Leggett	1	10	
" 3	do	Lt Leggett		30	
" 6	2674	Lt Leggett	1	15	Hooge
" 7	2674	Lt Leggett	1	30	do
" 7	2674	Lt Leggett		5	Aerodrome
" 7	F.E. 5643			15	do
	Total time		174	23	

Height.	Distance.	Weather.	Remarks.
7,000 ft		very cloudy	Bomb 100lb, to be dropped on railway line between Courtrai & Tourcoing, very cloudy, clouds at 1,000 ft. Flew due E, came down through clouds somewhere between Menin & Courtrai on River Lys, climbed over clouds again, flew easterly direction, came down through clouds again, but failed to get my bearings again, by this time it was practically dark. Decided to come home, flew due West, & came down through clouds over Bailleul, dropped bomb, with safty clip attached, in ploughed field. Flew round Bailleul until No:1. Squadron put out flares, & landed, 6-20 pm, quite dark.
		Normal	Flew machine back to Aerodrome
8000 ft		do	Photographs, one taken = I.12.a.0.4 at 5,200 ft, others at 7,000 to 8000 ft.
8000 ft		do	Photography, experimental.
5000 ft		do	Experimental bomb-dropping.
8000 ft		do	Photography
8000 ft		do	do
500 ft		do	Engine failure
2000 ft		do	First flight on F.E.2.b.

RECORDS OF FLIGHTS.

Date.	Machine.	Passenger.	Time in Air. Hours.	Time in Air. Minutes.	Course.
Time carried forward			174	23	
October 10	2674	Lt. Leggett		10	Hooge
" 11	2674	Lt. Leggett	1	0	do
" 23	5643	Lt. Griffin		40	
" 24	2674	Lt. Morgan		70	
" 24	2674	—		15	
" 26	5644	Lt. Morgan	2	0	do
" 30	5643	Lt. Griffin	2	25	
November 4	2674	Flt. Sgt. Bosworth	1	35	Hooge
Total time			184	38	

Height.	Distance.	Weather.	Remarks.
6,000 ft		Normal	Photography, Hooge, successful
5,000 ft		do	do very successful
7,000 ft		do	Patrol duty.
4,000 ft		Foggy.	Ranging. Landed at Ostyleterm behind the Belgium trenches, owing to fog.
800 ft		do	Flew machine back to Aerodrome
8000 ft		Normal	Artillery observation.
9000 ft		do	Patrol duty, attacked by enemy twin tractor & Aviatik. We dived down onto the Aviatik, which was 1,000 ft lower than we were, firing half a drum, when our gun jammed, at the same time I observed twin sticks, diving down on us, firing at about 500 yds range. We headed for him, which caused him to turn to his right, & away from us, at the same time, I turned sharply to the left, & put my nose down & flew over our own lines. Failing to repair gun, came home.
6,000 ft		do	Photography, while taking photos, witnessed duel between F.E. & Focker, saw Focker turn over on her back, & nose-dive, obviously out of control, & crash inside our lines about Zillebeke.

RECORDS OF FLIGHTS.

Date.	Machine.	Passenger.	Hours.	Minutes.	Course.
		Time carried forward	184	38	
November 7	2674	Yeo Draper		25	Aerodrome
" 7	2674	Lt. Howey		35	do
" 8	5644	Lt. Cave		20	do
" 11	2674	Flt. Sgt. Bosworth	1	20	Hooge
" 11	2674	Flt. Sgt. Bosworth		40	
		Total time for years 1913, 1914 & 1915	187	58	
		Total time for 1915	133	35	
		Total time for 1914	19	9	
		Total time for 1913	35	14	
		Total time overseas	132	15	
		Total time	187	58	

Height.	Distance.	Weather.	Remarks.
4000 ft		Normal	Testing machine.
3000 ft		do	Photography test.
4000 ft		do	Tuition
7000 ft		do	Hooge, Photography.
6000 ft		Cloudy.	Photography, too cloudy.

RECORDS OF FLIGHTS.

Date.	Machine.	Passenger.	Time in Air.		Course.
			Hours.	Minutes.	
	Time carried forward		187	58	
1916. January 12	H.F. 563	Capt. Chamberlin		10	Hounslow Aerodrome
" 12	do			10	do
" 14	V.F. 5655			10	do
" 14	BE2c 2085			15	do
" 15	do			10	do
" 16	2687	2/Lt Southhill		10	do
" 16	2092	Lt. Cam Duff		50	do
" 19	V.F. 5668			30	Farnboro' to Hounslow
" 20	BE2c 2092	Sgt. Trevis		20	Hounslow Aerodrome
" 20	do	Lt. Preston		15	do
" 27	D.H. Scout			15	C.F.S. Aerodrome
" 28	do			30	C.F.S. to Farnboro'
February 2	do			20	Farnboro' Aerodrome
" 5	do			88	Farnboro' to St Omer
" 20	BE2c 7325	Lt Evans	2	45	Farnboro' to St Omer
" 22	BE2c 7373	Lt Jones		20	Farnboro' Aerodrome
" 28	F.E.2.b. 6361	Lt Collar		90	Farnboro' to Folkestone
March 1	do	do		45	Folkestone Aerodrome
" 1	do	do		70	Folkestone to St Omer
" 9	BE2c 7332	2/Lt Sergeant	1	50	Farnboro' to Folkestone
" 10	do	do		20	Folkestone Aerodrome
" 14	do	do	1	5	Folkestone to Coquelles.
	Total time		203	16	

Height.	Distance.	Weather.	Remarks.
100 ft		Normal	First flight on H.F.
100 ft		do	
100 ft		do	First flight on Vickers' Fighter
500 ft		do	Engine missing.
500 ft		do	Test.
100 ft		do	Test.
4,000 ft		do	Test O.K.
3000 ft		do	Valve rod broke in the air & pierced rear spar.
2000 ft		do	Engine test.
1,000 ft		do	Joy ride.
1,000 ft		do	First flight on DeHavilland Scout.
300 ft		Very cloudy & rain	Flew at 200 to 300 ft owing to clouds & mist, very bumpy.
2000 ft		Normal	Testing.
5000 ft		do	From time leaving ground Farnboro', & landing at St Omer = 88 mins
8,000 ft		do	
3000 ft		do	Testing.
2000 ft		Stormy	Landed owing to snow storm.
1000 ft		do	Testing
6000 ft		do	
7,000 ft		Normal	Engine trouble
3000 ft		do	Test.
1,600 ft		Cloudy & fog.	Ran into fog, four attempts at landing, finished up in a dyke, smashing undercarriage & prop. (Coquelles 3 mls from Calais)

RECORDS OF FLIGHTS.

Date.	Machine.	Passenger.	Time in Air. Hours.	Time in Air. Minutes.	Course.
	Time carried forward		203	16	
March 19	F.E. ?	Sgt. Emery	5	0	Farnboro' to St Omer
" 24	B.E.2.c	Lt Budd	2	20	do
" 29	F.E.2.6 F.E.2.e	A.I.D. Tester	1	30	Farnborough Aerodrome
" 31	Martinsyde 1201P		1	30	Farnborough to St Omer
April 9	D.H. Scout 5961			35	Farnboro' test.
" 10	do			15	do
" 10	do		1	30	Farnboro' to St Omer
" 13	B.E.2.c			5	Farnboro' Aerodrome
August 10	B.E.2.c			20	do
" 20	F.E.2.d. 4926	Mr Henry Bremner	5	50	Renfrew (Glasgow) to Oxford
" 23	do	Mechanic	1	10	Oxford to Farnborough & test
" 24	F.E.2.6 ?	Gunner		15	Farnborough Aerodrome
" 24	do	do		55	Farnborough to Folkestone
" 24	do	do	1	0	Folkestone to St Omer & test.
" 30	BE12 = ?		1	30	Farnborough to St Omer
" 30	B.E.2.e = ?		1	0	St Omer to Candas
	Total time		228	1	

Height.	Distance.	Weather.	Remarks.
8000 ft		Normal	Shot raider down in mid-channel 8 mls N.E. of Deal (see report)
7,000 ft		do	"
5,000 ft		do	Comparative test, 120 H.P. & 160 H.P.
7,000 ft		do	Landed on rough ground at St Omer, broke axle & prop: while taxying, landing O.K.
		do	
		do	
7,000 ft		do	Forced landing ¼ mile from St Omer landing OK, restarted, flew to Aerodrome.
		Impossible	Testing machine, machine took charge owing to strong wind (gale lasted 4 days) machine crashed into gasometer R.A.7. buildings (see photos) damages, machine total wreck, myself broken collar bone & concussion also suffering from shock.
3000 ft		Normal	First flight after crash (3 months)
8000 ft		Cloudy & rain weather unsuitable for flying.	Forced landing, valve tappet broken, also petrol ran out, mechanic filled main petrol tank with 9 galls oil, disconnected same, & ran on front tank. made temporary tappet out of bolt, engine O.K.
3000 ft		Cloudy	Engine & machine OK. ran into rain storm.
1500 ft		Normal	
2000 ft		Strong wind	Strong N.W. Wind, very bumpy, record run on F.E.
3,000 ft		Low clouds	"
5,000 ft		Normal	
2000 ft		do	First trip on B.E.2.e.

RECORDS OF FLIGHTS.

Date.	Machine.	Passenger.	Time in Air. Hours.	Minutes.	Course.
1916.	Time carried forward		228	1	
September 2	Martinsyde Scout 160 H.P.			28	Farnborough to St Omer, & last
" 15	B.E.2.c		1	6	Farnborough to C.F.S. & last
" 19	F.E.8. Scout.			10	Gosport Aerodrome
" 20	Avro. 80 H.P			20	Cowes. Isle of White
" 28.	F.E.8.?			15	Aerodrome, Hurst Park
" 30	"			25	Hurst Park to Farnboro'
		Total.	230 hrs 45 mins		

Height.	Distance.	Weather.	Remarks.
		Very cloudy, rain	Forced landing, cam shaft broken,
300 ft		clouds 400 ft	Machine OK; landed near Guilford.
3000 ft		Normal	
2000 ft		do	First trip in F.E.8. Good landing
2000 ft		do	Inlet valve broken, forced landing machine OK.
1000 ft		Ground mist	Engine test, OK.
2000 ft		Cloudy, ground mist, very bumpy	Landed, owing to feeling ill in the air.

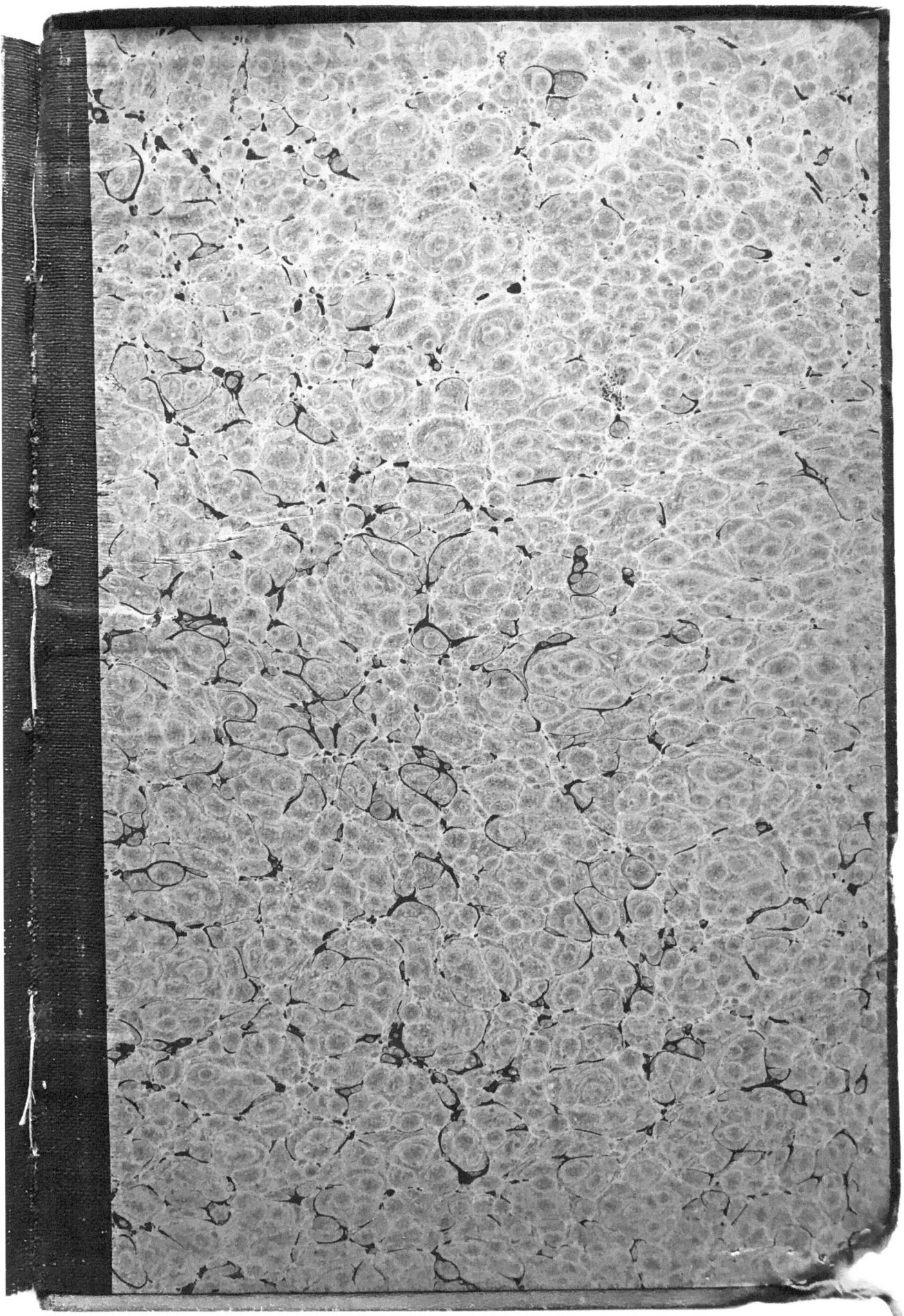

Canada

Reginald Collis continued to make a contribution to the war effort in passing on his engineering knowledge and experience when he was posted to No. 2 School of Aeronautics later that year, and from there to Canada to teach again at the No. 4 School of Military Aeronautics in Toronto, Ontario.

While still in Canada, Collis transferred from the Army to the RAF, after its formation on 1 April, 1918, continuing to serve as an engine instructor at No. 4 School until the end of the war.

Collis with fellow staff at the No. 4 School of Military Aeronautics (front row, second from right) note he is wearing his issue leather flying coat. *(Mark Hillier Collection)*

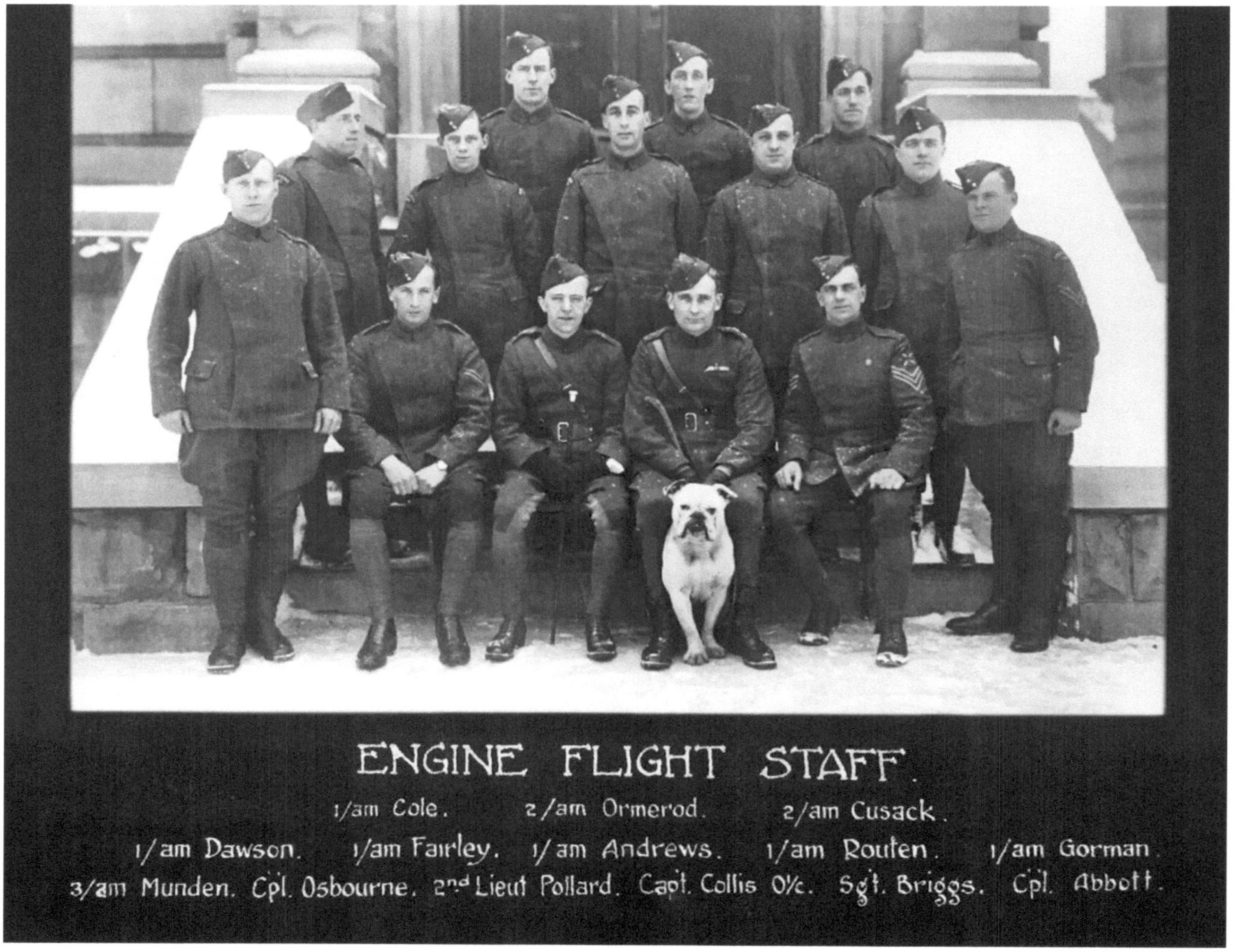

ENGINE FLIGHT STAFF.

1/am Cole.   2/am Ormerod.   2/am Cusack.
1/am Dawson.   1/am Fairley.   1/am Andrews.   1/am Routen.   1/am Gorman.
3/am Munden.   Cpl. Osbourne.   2nd Lieut Pollard.   Capt. Collis O/c.   Sgt. Briggs.   Cpl. Abbott.

The Engine Flight Staff at No 4 School, 1917. *(Mark Hillier Collection)*

Classroom instruction at the engineering section. *(Mick Prodger Collection)*

Above: Classroom instruction in airframes and rigging. *(Mick Prodger Collection)*
Below: Engine test sheds at No. 4 School of Military Aeronautics, Toronto. *(Mick Prodger Collection)*

Instruction in map-reading and navigation. *(Mick Prodger Collection)*

Returning home to England after the ceasefire, Collis went on the retired list in 1920, and found employment with an automobile engineering company in Burnley, Lancashire.

After three years, he discovered that civilian life wasn't for him, and in October 1923, he returned to Canada where he accepted a commission in the Canadian Air Force (later to become the Royal Canadian Air Force on 1 April 1924). The RCAF would become his full-time career. He was appointed officer in command of the ground instructional school at Camp Borden from 1924 until 1926, when he transferred to High River, Alberta as officer commanding No. 2 Operations Squadron. In November 1927 Collis was posted to No. 1 Repair Depot in Ottawa.

In 1930, he went to Montreal as resident A.I.D. inspector at Canadian Vickers, returning to Ottawa to become chief Trade Test Officer at RCAF headquarters later the same year, rising to Staff Officer training and finally Staff Officer in charge of inspection of aircraft.

Promotion to Squadron Leader followed in 1934 and in 1937 he was posted to Camp Borden to organize and command the No. 2 Technical Training School. Collis was promoted to wing commander in April 1938 and a year later became Engineer Staff Officer at No. 1 Training Command HQ in Toronto.

In September 1939, newly promoted to Group

Group Captain Collis in a rare moment of relaxation. *(Mick Prodger Collection)*

Captain, Collis was tasked with organizing and commanding the No. 1 Manning Pool in Toronto, where he remained until he took command of the Technical Training School in St. Thomas, Ontario in February 1940. This was the largest and most important techinical training centre under the British Commonwealth Air Training Plan, vital for supplying not just qualified, but exceptional pilots for the Allied war effort – at its time of greatest need.

In November 1941 Collis was posted to Headquarters and was responsible for the maintenace of all military aircraft in Canada.

After serving in two World Wars, Reginald Collis retired from the RCAF with the rank of Air Commodore.

Throughout his life he received many accolades for his achievements in engineering, both in military and civilian circles. In 1929 he was elected an associate member of the Royal Aeronautical Society and subsequently made an associate fellow. He was awarded patents for his designs and on 17 February 1936 was granted membership of Chartered Institute of American Inventors.

The leather flying coat issued to and worn by Reginald Collis during his days of aerial combat with the RFC. It was obtained directly from Collis's family by a private collector along with the French belt shown. It is now held in a private collection in Canada. *(Courtesy Tony Schnurr/Kaiser's Bunker)*

Combat Flying Timeline

1909	Blackpool airport, maintaining Bleriot monoplane.
26/06/12	After originally enlisting as a Sergeant in the 2nd East Surrey Regiment, Collis is seconded to the RFC as an Air Mechanic 1st Class and given the RFC number 109.
29/01/13	Collis flies 40 minutes with a passenger (one of the final requirements before taking his test for the RAeC Aviator's Certificate).
04/02/13	Collis passes his RAeC test in a Maurice Farman, having trained at the Central Flying School, Upavon, receiving certificate number 412.
2/11/14	Promoted Sergeant at CFS.
8/3/15	Graded First-Class flyer.
10/5/15	First flight in France at RFC HQ, St Omer.
16/05/15	Moves from St Omer to La Gorgue Airfield, Home of 16 Squadron.
29/06/15	Collis is commissioned as a 2nd Lieutenant in the field for services rendered.
16/07/15	Posted to 6 Squadron, based at Abeele, on the border of Belgium and France.
24/11/15	Collis suffers a 'debility' and recovers in a convalescent home until to 18/12/15.
January 1916	Posted to 24 Squadron at Hounslow.
22/01/16	Posted to 1 Reserve School for delivery of machines to the front.
16/09/16	Posted 41 Squadron forming at Gosport with the F.E.8. He is there one month before becoming unwell and does not fly again.
	Posted No. 2 School of Aeronautics.
	Posted No. 4 School of Military Aeronautics, Toronto, Canada.

Bibliography & Online Sources

Campbell, G.L., assisted by Blinkhorn, R. H., *Royal Flying Corps (Military Wing) Casualties and Honours During the Great War 1914-1917,* London, 1917

Cole, Christopher, (edited by), *Royal Flying Corps Communiqués 1915-1916,* William Kimber, 1968

Comini, Serge, *1914-1918 - Abbaye Notre-Dame de Beaupré-sur-la-Lys, des Hommes sur les terres et les chemins des Dames,* September 2018

Cooksley, Peter, G., *The Royal Flying Corps 1914-1918,* Spellmount, Gloucestershire, 2004

Grinnell-Milne, Duncan, *Wind in the Wires,* Hurst & Blackett Ltd, London, 1933

Hawker, Tyrrell, *Hawker VC, RFC Ace, The Life of Major Lanoe Hawker VC, DSO 1890-1916,* Pen and Sword, 2013

Henshaw, Trevor, *The Sky Their Battlefield,* Grub Street, London, 1995

Hobson, Chris, *Airmen Who Died in the Great War,* J. B. Hayward & Son, 1995

Jefford, C.G., *Observers and Navigators and Other Non-Pilot Aircrew in the RFC and RNAS,* Grub Street, London, 2014

Johnson, Steve Buster, *Over the Western Front, 6 Squadron Royal Flying Corps,* FeedARead.com, 2018

Lewis, Cecil, *Sagittarius Rising,* Folio Society, London, 1998

Longmore, Arthur, *From Sea to Sky,* Geoffrey Bles, London, 1946

Macmillan, Norman, *Into the Blue,* Duckworth, London, 1929

McInnes, I. & Webb, J.V., *A Contemptible Little Flying Corps,* London Stamp Exchange, London, 1991

McCudden, James T.B., Flying Fury, *Five Years in the Royal Flying Corps,* Bailey Brothers and Swinfen Ltd, 1973

Rawlings, John, *Fighter Squadrons of the RAF and Their Aircraft,* MacDonald, London, 1969

Shores, Christopher, Franks, Norman, Guest, Russell, *Above the Trenches,* Grub Street, London, 1990

Spinks, *Medal Circular,* Issue No.13, August 1999

Strange, L.A., *Recollections of an Airman,* Hamilton, London, 1935

Taylor, W.R., *CFS Birthplace of Air Power,* Puttnam, London, 1958

Williamson, H.J., *The Roll of Honour RFC and RAF 1914-1918,* Naval and Military Press, Dallington, 1992

www.airhistory.org.uk

www.iancastlezeppelin.co.uk

www.stevebusterjohnson.com